An Antarctic Summer in 1968-69

Tide Cracks and Sastrugi

TIDE CRACKS AND SASTRUGI
AN ANTARCTIC SUMMER IN 1968-69

Graeme Connell

To Ken
Enjoy the journey!!
Graeme
Nov 2013

Queen Maud Land

Weddell Sea

South Pole
Amundsen-Scott (US)

Ross Ice
Shelf

Scott Base

Vanda Station

McMurdo
Sound

**Ross
Sea**

RECOGNITION OF MY ROOTS

This is a tribute to my beloved portable typewriter, a Hermes 3000 of the late '60s. The many stories we created together helped me make a living.

While I now live in North America, I chose to create Tide Cracks and Sastrugi in the language and terminology of the day, hence the English spellings and Imperial measures.

I've relied on my own resources for much of the book with loads of help from family and friends. My own Antarctic library has enabled me to check facts and verify data. I am grateful for these accounts. Any mistakes are purely mine.

Marcello Manzoni gave me the wonderful picture for the cover, and Lois bravely drew the maps. I took all the photographs except where credited.

And thanks Hermes, you remind my grandchildren that you were around before computers.

*** *** ***

Tide Cracks and Sastrugi: An Antarctic Summer in 1968-69

E-Book ISBN: 978-0-9876922-1-4
Paperback ISBN: 978-0-9876922-0-7

Additional copies of this book may be ordered by visiting the
PPG Online Bookstore at:

🍁**PolishedPublishingGroup**

shop.polishedpublishinggroup.com

Due to the dynamic nature of the Internet, any website addresses mentioned within this book may have been changed or discontinued since publication.

FOR

Lois
Hilary, Rachel, Bridget
Max, Robert, Veronica, Fisher, Emma, Ethel, Beth, Gillian.

Where am I going? I don't quite know.
What does it matter where people go?
Down to the wood where the bluebells grow –
Anywhere, anywhere. I don't know.

--- A.A. Milne, Spring Morning.

What's Inside

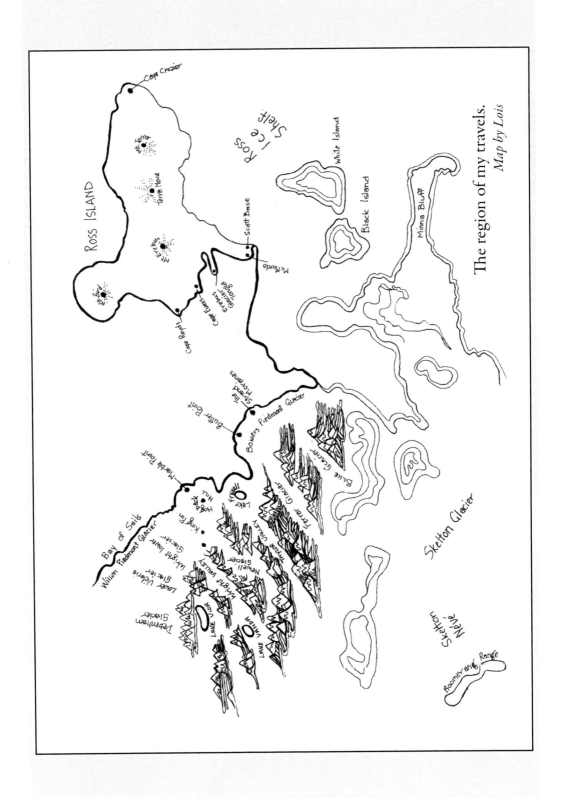

The region of my travels.
Map by Lois

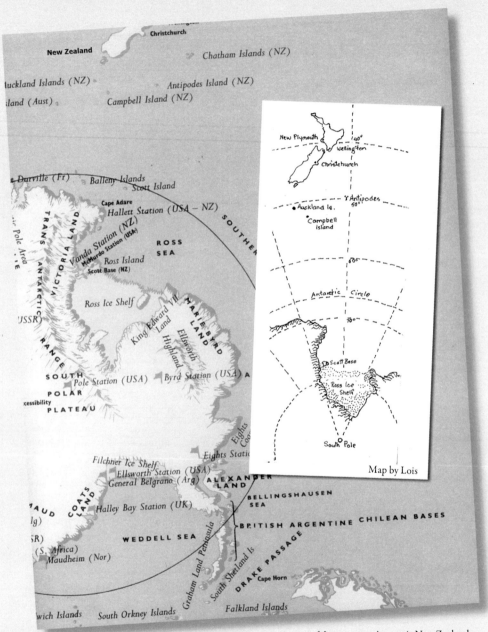

New Zealand

Christchurch

Chatham Islands (NZ)

Auckland Islands (NZ)

Antipodes Island (NZ)

Island (Aust)

Campbell Island (NZ)

Durville (Fr)   Balleny Islands
                                    Scott Island
                    Cape Adare
                    Hallett Station (USA – NZ)
            Vanda Station (NZ)
            McMurdo Station (USA)   ROSS
                    Ross Island   SEA
                Scott Base (NZ)

Pole Area

T R A N S   V I C T O R I A   L A N D

Ross Ice Shelf

(USSR)

King Edward VII Land

M A R I E · B Y R D   L A N D

Ellsworth Highland

A N T A R C T I C

R A N G E

SOUTH

Pole Station (USA)   Byrd Station (USA)   A

POLAR

cessibility

PLATEAU

Eights

Filchner Ice Shelf
Ellsworth Station (USA)
General Belgrano (Arg)   ALEXANDER
                          LAND

Eights Statio

BELLINGSHAUSEN
SEA

MAUD   COATS
lg)   LAND

Halley Bay Station (UK)

BRITISH ARGENTINE CHILEAN BASES

SR)
(S. Africa)
Maudheim (Nor)   WEDDELL SEA

Graham Land Peninsula

South Shetland Is.

DRAKE PASSAGE

Cape Horn

wich Islands   South Orkney Islands   Falkland Islands

New Plymouth   40°
            Wellington
              Christchurch

Antipodes
                          50°
Auckland Is.
Campbell
Island

60°

Antarctic   Circle

70°

Scott Base
          Ross Ice
          Shelf

South Pole

Map by Lois

Map courtesy Antarctic New Zealand

The jewel of the McMurdo Dry Valleys, Lake Vanda, deep in the heart of the Wright Valley.

Back from the field, members of the Victoria University of Wellington 13 expedition enjoy an evening at Scott Base before heading home to New Zealand. From left: Barrie McKelvey, Barry Kohn, Peter Webb and Mike Gorton.

FOREWORD

Antarctic Science and Exploration

As Antarcticans reflect on the details and meaning of their polar experiences, they can't but help relate them in every way to the rest of the journey through life, specifically, one's origins, family, friends, colleagues, mentors, forks in the road, unexpected opportunities, as well as the enlightened, unfortunate and dubious decisions we all make. We acknowledge then, that the Antarctic experience had a fundamental influence on our lives. Even for those of us fortunate enough to have visited Antarctica many times, the first visit to the continent remains unique in so many ways, and Tide Cracks and Sastrugi: An Antarctic Summer in 1968-69 helps us recapture the sheer exhilaration and wonderment of that first visit.

The New Zealand Antarctic Programme (NZARP, later NZAP) embraces a remarkable history spanning more than 50 years. Many official expedition and personal accounts have been published and thousands of scholarly publications attest to the success of its diverse science programmes. All these contribute to the cumulative Antarctic historical record. So as to offer a little historical perspective to Graeme's account, I pose this question. What was unique about the NZARP 1968-69 Scott Base and field campaign and where does it fit into a half century of New Zealand Antarctic history? Only a decade earlier, the New Zealand elements of the TransAntarctic Expedition (TAE), International Geophysical Year (IGY), and International Geophysical Cooperation (IGC) programmes (1957-59) inaugurated New Zealand's independent entry into Antarctic exploration and science. NZARP 1968-69 exhibited many ghosts of these formative years, while at the same time demonstrating an urge to strike out in new scientific, logistical, and technological directions. In 1968-69 Scott Base was essentially the same place as it was a decade earlier, except that it was now painted green rather than yellow! TAE/IGY era over-snow transport relics such as a Sno-Cat, Weasels (World War 11 era tracked military vehicle designed for snow) and a NZ Army Bren gun carrier (the engine being used for the Scott Base ski tow) were still in use. NZARP continued to rely heavily on the National Science Foundation (NSF) /US Navy for logistic support both between Christchurch, New Zealand, and Antarctica, and to distant field study areas in the Transantarctic Mountains and other remote regions. McMurdo Station still boasted many grubby temporary buildings from Operation Deep Freeze a decade earlier. Although US Navy aircraft such as ski-equipped Hercules C-130 and new generation helicopters had been added to the air armada, two 1953-vintage Super Constellations'(C-121) (Pegasus and Phoenix) still plied the air route between Christchurch and McMurdo Sound. Communications were basic and often unreliable, and Morse code was still a language often used in field communications within Antarctica and between Scott Base and New Zealand. Mechanical typewriters were still in fashion; roll film cameras captured images, and colour slide film had to be sent to Australia for processing. Weeks-old New Zealand newspapers were eagerly read at Scott Base. Old style gramophone records provided music and an archaic Bell and Howell projector chattered forth a constant diet of the best that Hollywood had to offer. Regular mail service depended on unscheduled air links to the outside world; fellows in remote field parties might receive only two or three letters in a whole season, and by then the news was usually well out of date. The age of computers, complex software, email and digital photography still lay a decade or two in the future. Today, through the miracles of real time email and satellite communication, it is possible for the intrepid Antarctic explorer, even at a remote field site, to learn several hours after the event, that back home, the car just exploded, the plumbing suffered a catastrophic failure, that the bank had just re-possessed the house, and that several of the more valuable children were still missing and presumed lost! We also tend to forget that in the pioneer days of the Trans-Antarctic Expedition and International Geophysical Year (1957-58), New Zealand's manpower with real polar experience numbered only a few tens of people and that a decade later in 1968-69 numbers had ballooned to several hundred. Such was the 1968-69 polar experience.

New Zealand's very active role in Antarctic science and exploration during the 1960's was administered through the Ross Dependency Research Committee and National Committee for Antarctica. The country also participated actively in the International Committee Of Scientific Unions (ICSU) and Scientific Committee On Antarctic Research (SCAR), as well as the Antarctic Treaty organization. The Department of Scientific and Industrial Research (DSIR) and its subgroups, including Antarctic Division, NZ Geological Survey, Geophysics Division, Soil Bureau, and the NZ Lands and Survey Department, continued to actively sponsor scientific programmes in Antarctica. Victoria University of Wellington and University of Canterbury, Christchurch, contributed to the scientific effort in earth sciences and biology respectively and facilitated the emergence of the next generation of young Kiwi polar scientists.

The 1960s were particularly interesting and challenging years for the Antarctic geologist. The reconnaissance geological and geophysical surveys accomplished during TAE-IGY-IGC produced interesting results but, as geologists, we had hardly scratched the surface of the earth science bonanza that awaited future workers. By 1960, the Continental Drift hypothesis espoused by Alfred Wegener in the first decade of the 1900's to explain continental asssembly of the various fragments of Austrian geologist Eduard Suess' Gondwana supercontinent, had at last been accepted by the majority of earth scientists. The arrival of companion hypotheses on sea floor spreading and plate tectonics in the early 1960's reinforced the central role of Antarctica in comprehending global geological events over scales of hundreds of millions of years. Through the 1960's, and including the earth science programmes of the 1968-69 season, geologists continued to build data bases in stratigraphy, tectonics, paleontology, petrology, geochronology, and paleomagnetism, etc, that allowed better understanding of Antarctica's pivotal keystone role in refining the Gondwanaland continental assembly with India, Africa, Australia, South America, and New Zealand, and also to comprehend the Gondwanaland (including Antarctica) relationship with Laurasia (another super-continent) and Pangaea (the combined Gondwana-Laurasian mega-continent). By 1970 New Zealand government and university scientists had made major contributions to these goals via their fieldwork along the length and breadth of the Transantarctic Mountains.

Although Proterozoic, Paleozoic and Mesozoic (rocks older than 65 million years) earth science continued to undergo more detailed investigation and refinement in the decades from 1970, attention now turned to the later Antarctic earth history (the Cenozoic or last 65 million years), and deciphering the complex history of Antarctica's long terrestrial glacial and ice sheet record. The Dry Valleys and other regions of the Transantarctic Mountains became focal points in this endeavour. The establishment of New Zealand's Vanda Station in the Wright Valley during the 1968-69 summer provided an essential logistic centre from which to launch the many new science programmes directed towards investigation of Cenozoic glacial history. In Tidecracks and Sastrugi we read of the considerable logistic effort to establish Vanda Station on the eastern shores of Lake Vanda. While conventional geological exposure fieldwork on the glacial record continued unabated , only two to three years later in 1971-72, a technological threshold loomed with the inception of deep geological drilling in Taylor, Wright and Victoria Dry Valleys. Vanda Station provided an important logistic center for operations by the Dry Valley Drilling Project (DVDP), a multi-year project sponsored by the United States, New Zealand and Japan.

Tidecracks and Sastrugi adds valuable new detail and insights to the steadily mounting flow of historical information on New Zealand's role in Antarctic exploration and science over the past half century. One of the great successes of NZARP over the years is the degree to which the New Zealand public and politicians have been kept so exceptionally well informed and educated by its media representatives in the field. New Zealanders must be the best informed on polar matters among the many Antarctic Treaty nations. As we see in the pages that follow, Graeme made a major contribution to this end during the time he spent in Antarctica as the NZ Press Association correspondent. We acknowledge and savour his ability to bring forth the fascinating minutiae of daily life at base and in the field and the significance and successes of the science endeavour, all information that makes Antarctica such an interesting, challenging and unique environment in which to live, work, grow and ponder. No national polar science program achieves its lofty goals without the contribution of its enthusiastic, talented and creative cast of science support personnel, those 'genius' mechanics, carpenters, cooks, electricians, tractor drivers, field assistants, and base leaders. This book acknowledges warmly their indispensable toil and contributions.

While the 1960s era male explorers indulged their boyhood fantasies in the far south, parents, spouses and children remained behind to cope with the daily grind in the normal world. The variety of personal emotions expressed herein, with exchanges between separated family members, feature prominently in this account. A baring of the soul is an uncommon feature in Antarctic literature. If we are really honest, these complex interpersonal issues are something all Antarctic veterans experienced at some point. We applaud Graeme's courage in reminding us all that life on the home-front goes on regardless and should be acknowledged.

Although I participated with Graeme in NZARP during the austral summer of 1968-69 we meet again some 43 years later within the pages of this book. I'm delighted to be invited to provide this Foreword which has allowed us both a chance to recall, reflect on and share events that occurred during a very productive and enjoyable field season long, long ago. As is the case with most Antarctic veterans, life has taken us in very diverse and challenging directions. Today, we find ourselves residing on the same continent: Graeme in Alberta, Canada, and me in Ohio, USA. Our early life had much in common. We experienced the pleasures of rural and coastal city life in distant New Zealand, lived in historic and picturesque Taranaki province, attended New Plymouth Boys' High School, learned from the same teachers, and shared many of the same city and farming families and friends. Images of life in the 1950's and 1960's New Zealand were easily rekindled as I penned this Foreword. I do not recall discussing these common roots and experiences during our time in Antarctica in the summer of 1968-69. Strange perhaps, but herein lies a point worth making. Our perspective and understanding of life's journeys are perhaps more objective and insightful given a long passage of intervening time. It is with the thought that a single important event in life, such as going to the Antarctic, provides a convenient stage upon which to dwell on what came before and after this memorable experience.

Peter-Noel Webb
School of Earth Sciences
The Ohio State University
Columbus, Ohio, USA

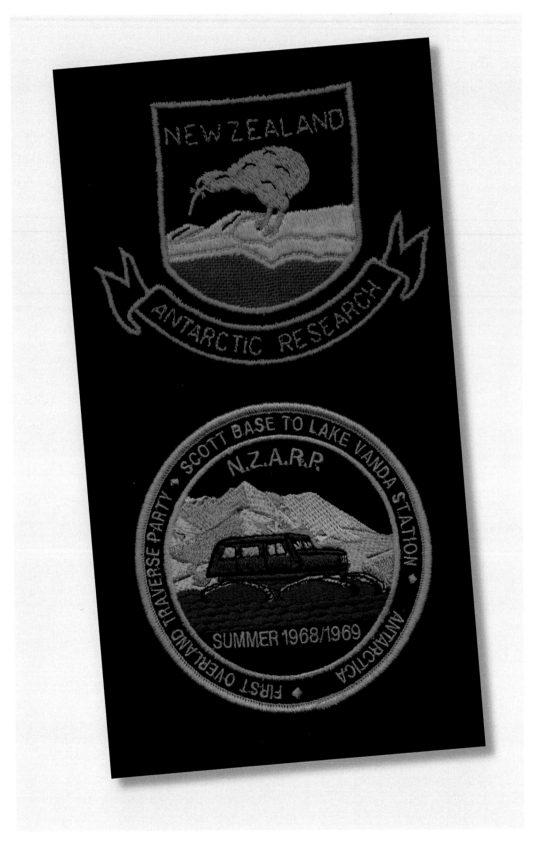

INTRODUCTION

"Plucked From The Comforts Of Home"

The New Zealand Antarctic Research Programme's main thrust in the summer of 1968-1969 was the establishment of a winter-over base at Lake Vanda in the dry valleys of Victoria Land, Antarctica. To support this we needed to transfer building materials and supplies over the surface using tractors towing sledges, as there was no extra helicopter time available. Graeme with no knowledge of travel in Antarctic conditions was plucked from behind his reporter's desk and thrust into the unfamiliar role of field assistant on the Vanda tractor train travelling from Scott Base, over the sea ice of McMurdo Sound and up on to the Wilson Piedmont Glacier which formed a barrier to the entrance to the Wright Valley. We were short of field assistants and we anticipated that it would be a good thing if Graeme could be on hand to send out progress reports to the New Zealand Press Association. Good reporting of New Zealand's Antarctic activities by word (radio and Morse code) was extremely important back in those days. Antarctic Division funding was, in a sense, dependent on the man on the street through his taxes, actually paying all of us to enjoy, probably the number one adventure pastime available to New Zealand males. Therefore, so that every New Zealander could be given the opportunity to see how his/her money was being spent, it was necessary to have solid media coverage throughout the summer season.

It was of paramount importance to our scientists to have a team winter over at the new base. It was also important for our group to establish the new base at no extra cost to the Government. The science community wanted to probe deeper into why the dry valleys were snow-free and why, for instance, Lake Vanda's bottom temperature was more or less equal to that of bath water.

And so it was that Graeme joined several other untried young men, plucked from the comforts of their homes back in New Zealand to drive a hotchpotch of worn out machinery across a frozen sea, up and over a crevassed area and on to the floor of the Wright Valley. It is with some unease that I, as leader of NZARP 1968/69, recall this event. Some planning went into the venture, but by today's standards it would be deemed inadequate. For instance, I did not give the group written instructions. After an earlier on site reconnaissance there was general consensus amongst the leadership that a route down into the middle of the valley would offer the best chance of success to get from the piedmont to the valley floor. However, a field decision was made to take the largest two vehicles into a dangerous area resulting in damage to both. I'm sure the extrication of these vehicles from crevasses will make some great reading and as often happens, setbacks and the correcting of them results in the development of a camaraderie, that otherwise may not have been born. This mateship is alive and well, to this day.

Within the past year Graeme has confided in me that he began to be aware of God while in Antarctica. He tells me his interest was heightened through an unwitting action by me. More as an attempt to find another activity for those staying at Scott Base, I saw attendance at the Sunday morning services at the Chapel of the Snows at McMurdo Station as a worthy alternative to sleeping in. I took a regular group 'over the hill' to hear Padre Harold Baar, US Navy, a Lutheran minister. Members of the Scott Base team became an integral part of those services. I often read one of the lessons and Chris Rickards, our electrician, played the organ. Graeme became aware of God when he (and his companions) almost died while cooking their dinner one night during the traverse to the dry valleys. Shortly after his near death experience he found himself speaking to God, promising that he would be a good husband and father if he survived his Antarctic travails. To my knowledge he has kept his faith and I applaud him for that.

The rest of Graeme's summer was spent reporting NZARP 1968/69's endeavours, the comings and goings of dignitaries that included the Governor

General of New Zealand, His Excellency Sir Arthur Porritt. He extended his journalistic abilities by writing a base newspaper to keep us up to date with local and world affairs. He and two of the other summer support party, Alister (Taffy) Ayres and John (Chippy) Newman were the life and soul of the base and provided us all with some uproariously, hilarious moments. But I'm sure all of this will not beat the experiences he encountered on the Vanda Tractor Train.

Just last southern summer Graeme and I took an extended walk to the Muriwai gannet colony. Making our way along the beach and up the cliffs to the gannets, we soon fell into an easy and enjoyable reminiscence, recalling events and experiences from, not quite our barefoot days, but those of our mukluk-booted days of yore.

Robin Foubister,
Leader, NZARP 1968-69
Muriwai Beach
Auckland, New Zealand
Easter 2011.

The cover my dad made up for me at the **Taranaki Herald**
as the front page for my photo albums.

New Zealand Antarctic Research
Programme

SUMMER   1968-69

—— ★ ——

Information Officer - Photographer

GRAEME K. CONNELL

PROLOGUE

I still "see" our daughters (then) and the grandchildren (now) huddled over, peering into a rock pool and overhear them... "I wonder if...I wonder why..." Or it might be a rock with 'things' in it, perhaps a flower or a leaf. Next thing I'd know they are jumping on me looking for an answer. Thankfully I have the science types who have gone before to help me with good responses. While beakers and bunsen burners were a failing subject for me in four years at high school, 10 years as a daily newspaper journalist provided insight into why tuna might or might not swim off the Taranaki coast, the dairy industry of my famous province was more than milking cows, and that oil and gas exploration was more than poking a hole in the ground. In other words, practical experience showed me there's a science for everything.

Taking a job as a journalist with the New Zealand Antarctic Research Programme (NZARP) for the summer of 1968-69 parachuted me into the world of science and scientific endeavours in the remarkable frozen laboratory of Antarctica. As any of my New Plymouth Boys' High School teachers can attest I was not into science at any level way back then but in my writing career I have certainly enjoyed meeting plenty of scientists, talking with them and writing about their pursuits. I am very fortunate that throughout my careers as a journalist and a communication specialist in the oil and gas industries, to have worked alongside patient and knowledgeable people always ready to share their world. I admire their knowledge and I'm glad for their tenacity and interest in always wanting to learn more.

A team from Captain Robert Falcon Scott's Discovery expedition (1901-1904) was probably the first to get a look at the remarkable area of Antarctica known as the Dry Valley Region on the western shores of McMurdo Sound.

The valleys are 'dry' simply because there is no snow and ice (except on the lakes). It is a barren desert between equally barren and 'dry' mountain ranges. Sir Ernest Shackleton's western party during the 1908-1909 Nimrod expedition followed up on Scott's early exploration.

The Australian Griffith Taylor (after whom one valley is named), Frank Debenham and the Canadian Charles (Silas) Wright investigated the valley more thoroughly in their fieldwork during Scott's Terra Nova expedition in 1911-12. After examining the Taylor Valley the scientists were left wondering what lay to the north, between the Wilson Piedmont skirting the Ross Sea coast and the ice plateau of the hinterland. Next, along came the US polar explorer Admiral Byrd and his group. They took some aerial pictures of the region around 1947. Fast-forward and the International Geophysical Year and the Trans Antarctic Expedition of 1957-58 stimulated greater interest in this part of Antarctica.

In the summer of 1956-57, during preparations for the International Geophysical Year, a couple of curious Victoria University of Wellington geology students, Peter Webb and Barrie McKelvey, badgered the powers that be until they were allowed to travel south as cargo handlers on the New Zealand supply ship HMNZS Endeavour. With a mixture of perseverance and good luck the pair reached the area now known as the Victoria Valley complex. They returned the following year under the leadership of the University's Dr Colin Bull to make a more thorough investigation of the Wright Valley (the team created the name) lying between the Taylor and Victoria (after the university) valley systems.

The Dry Valleys gained international attention as their uniqueness became more widely known and in the years that followed Victoria University spearheaded numerous investigations into the region's geology, meteorology, physics, chemistry, seismology, gravity and biology.

You can't help but get carried away with the stuff these science guys have found out about the Wright Valley that shows evidence of four glaciations, the first two of which were the most intense. Both carried ice from the inland plateau on to McMurdo Sound and the Ross Sea. Highly weathered moraines of the first glaciation lie on glacial benches at heights of about 4500 feet above sea level. The second glaciation is a thin veneer of highly weathered moraine on the floor of the present valley and bench with remnants up to 200 feet above sea level. Thick moraines of the third glaciation partly overlie the second, indicating a less extensive advance into the valleys by plateau, alpine and coastal piedmont glaciers.

Puzzled investigators also paid a lot of attention to the lakes because the water was warmer deeper down than at the surface. In 1961-62 the water was recorded at 27°C 200 feet below a 12-foot thick ice cover. This created a bit of turbulence in the scientific world as investigators sought to find an explanation. Another scientific fascination was the discovery of mummified seal carcasses on the valley floor up to 45 miles inland. The 1958-59 team actually found 99 of them. By the mid-60s, it was deemed essential for a team of scientists to winter in the area to probe the region's many mysteries. Plans were made for a joint party of scientists from the US, Japan and New Zealand to winter over during the 1965 season at a base to be erected in the Wright Dry Valley. This fell through largely because of the already heavy US logistical and financial commitment and involvement in other projects. Eventually New Zealand's Ross Sea Dependency Committee agreed that the Antarctic Division of the Department of Scientific and Industrial Research should make itself responsible for a winter-over station.

The New Zealand interests decided a satisfactory station could be established at minimum cost by transporting several existing but disused huts to the Wright Valley and re-erecting them on a site near Lake Vanda. These huts included the auroral radar hut at Arrival Heights on Ross Island and the biological hut from Cape Royds, also on Ross Island. Economics entered the picture again in 1967 and plans were cancelled. A review showed that the mission could be partially accomplished in 1967-68 with a low cost summer base for use by a Victoria University party.

From those beginnings, the base would be upgraded during 1968-69 term to accommodate the first winter-over party in this intriguing and puzzling part of our planet. The thrust of a smaller than usual NZARP team, was to construct the country's first winter-over base on mainland Antarctica. Once in place, five men would spend a very lonely and isolated six months experiencing and recording winter's effect.

A fair number of "ologies" were involved in the projects detailed for winter study. The thrill of being the first New Zealand base on the continent of Antarctica meant the station would be manned by folk keen to find out stuff about meteorology of course but also about glaciology, pedology (soils), and hydrology, which all rely on good climatological data. Other projects listed included geology, seismology, upper air physics, and human physiology, as well as aurora and geomagnetic studies.

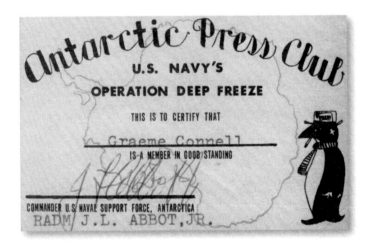

Even though this book is really about Antarctica, I just had to include a picture of Mt Taranaki (formerly Mt Egmont), the 8260 ft volcanic peak that is central to my home province of Taranaki. So here it is!

Wearing Antarctic-issue clothing, I negotiate the thin snow covering at 4500 feet on Mt Taranaki, the backdrop to our hometown of New Plymouth.

(Rob Tucker photo)

I.

LOOKING BACK

Perhaps my taste for adventure started with a bicycle.
In a letter to his son Eduard, Albert Einstein advised:
"Life is like riding a bicycle. To keep your balance you must keep moving."

Bicycles were my early passports to adventure and an instrument in my constant search for who knows what. Along the way there were a number of years when the bike lay forgotten, swapped from two wheels for four, one seat for five. But the bike came back: smoother, lighter and flexible.

The magic time of morphing from three wheels to two was accomplished using whatever bikes were around our place: sometimes my dad's and sometimes my elder brother's or sister's. The big trouble with these machines was they were too big for me so on the boy's bikes I used to put one leg through and under the top bar and kinda pedal sideways. It was a real struggle to balance and stay upright and the calf of my right leg was always sporting a greasy chain mark, often mixed with a bit of blood. When I decided to try and use the over the top bar method, my younger brother and I used the front hedge to climb on to the bike and push off to gain enough momentum to carry us across the five-foot gap at the front gate and see how far we could travel before toppling, hopefully, back into the hedge. Take off day did come and seated uncomfortably on the top bar, I swept thrillingly over the daisy-laden front lawn before daring to slip through a narrow pathway between garden shrubbery and the red corrugated iron garage. I wobbled midway through and caught my knee on the raw edge of metal. In agony I made it through and smashed head first over the bars and into a Macrocarpa tree (Monterey

Cypress). Blood poured from my knee as I lay in the tangled wreckage of bike and branch. I was ecstatic. I was now a cyclist!

After that I was forever "borrowing" my elder brother's bike, even my sister's, or Dad's. But by the time I graduated with parental approval to ride outside the gate I had a bike, a bitser, put together from scrounged or donated parts, one speed, foot brake only, black and rusty. But to me, it could fly, straight out the gate on to the gravel road—whoosh, it flipped out from under me. I sat in the sun and nursed my wounds, picking bits of stone and sand from my leg, knee, elbow and arm, washed in the tears of an over-excited pedaller, hoping all the time that my mother could not see me out her sewing room window.

The bicycle led to employment with a paper route. That led to income and that led to upgraded bikes and the cash needed for calliper brakes and a three speed. With my younger brother and friends we'd cycle to the river swimming holes, we zapped through the magnificent gravel pathways in the city garden parks hoping we wouldn't get caught. The bike took me to sandy beaches and the community swimming pool. I lived out my fantasies on those two wheels. One of my early purchases was a used machine that needed some work. I took it apart, scrubbed it down painted it a crazy pink, and had to get my dad to help put it back together. I dreamed up long distance events and coerced friends into going along with me.

My mother, a writer, really wanted her six children (I was No 3) to grow up as readers and if family funds allowed we could count on a book each Christmas or birthday. We were still some 20 years away from television. Books discovered at the end of my bed on Christmas morning were packed with stories of thrilling exploits: The Children's Own Treasure Annual, Collins Magazine, and Boy Scouts Annual. I supplemented these with library books like Enid Blyton's Famous Five and Arthur Ransome's Swallows and Amazons. The camping, sailing, fishing, exploring, and wild imaginations of kids my age supported my unsettled and unfocussed wannabee character.

The bike gave me freedom.

The books gave me dreams.

Then for Christmas of 1952, just a couple of months into my 12th year,

I ripped at the wrapping to uncover the latest Collins Annual. Through the smooth shiny black and white pages I found the article of articles, White Continent, by Robert Jackson who began his story:

"Turn a globe upside down and you will find yourself looking at the last of the world's great mysterious landmasses. . . the five million square miles of frozen, uninhabited, inhospitable land we call Antarctica."

Now I saw true adventure, way beyond the beaches and the ocean-going freighters of my hometown, way beyond Mt Taranaki (Mt Egmont) that 8260 ft near-perfect cone of a volcano we lived under. Antarctica was a continent of new mystery to my young Mittyesque mind and beat out darkest Africa as the place of ultimate adventure. I read the story and shelved the book, went outside and took off for a ride on my bike.

I blundered into 1953, an energetic boy scout badgering his scoutmaster to organize camping weekends where I could play in the river, build bridges out of ropes and sticks, cook fun meals on my own, hike, and plan new and exciting expeditions. It was my final year before high school. My income came from an after school paper route, and spare hours were filled with field hockey through the winter and swimming during the summer with club competition each Monday night. Boyhood was passing. I felt like I was in a hurry, always seeking a way of escape, to dodge the shadows that had settled over our home as my mother battled to survive her cancer and most of all the treatment. I think we were a rowdy lot, four boys and two girls (the bookends) and often, after the evening meal and the usual chaos and arguments around whose turn to wash and dry the dishes, we would hang out around the radio and listen to news and comedy programmes. It was good during the winter as the dining/living room was the place to be to save electricity and enjoy the warmth around a kerosene heater. Together we laughed and learned the wonders of words through the BBC radio shows of the 50s. "Take it From Here", "My Word!", "Hancock's Half Hour" and "The Goon Show" were the staples of the time. Words and language became life.

And that's where I was the night the radio crackled out the news of Edmund Hillary and Sherpa Tenzing topping Mt Everest with a British

Expedition. A New Zealander who at one point in his life had used our mountain as a training ground during the Second World War had conquered the world's highest peak. The poor reception and remoteness of the relayed BBC broadcast only served to heighten my imagination of being up there. Creating a scrapbook of media miscellany (a popular pastime with everybody but me around that time) was furthest from my mind. Something in this global adventure captured my attention though and I went out and bought a blank book and collected all the news items and media pictures I could about the epic climb. The one picture that stood out was that of Tenzing on the very summit of Everest holding aloft a flag-bedecked ice axe for Hillary to take the historic picture. A knighthood was conferred on Hillary and a month after the event, he told a correspondent that: "It felt damn good at the top. It was a beautiful day with a moderate wind. As we got there my companion threw his arms round me and embraced me."

This was world shattering news on the eve of the coronation of Queen Elizabeth who was expected in New Zealand at the end of the year on the first visit by a reigning monarch. Sadly, the day after her arrival with Prince Phillip, New Zealand experienced one of the world's largest rail disasters. All together 151 people died in that Christmas Eve tragedy when a natural ash dam broke at the crater lake of Mt Ruapehu sending a lahar (volcanic mudflow), a 20-foot high wall of water, ice, mud and rocks, down the Whangaehu River to smash into the concrete pylons of a railway bridge at Tangiwai (meaning weeping waters) just as a locomotive pulling nine carriages and two vans at more than 40 miles an hour reached the severely weakened bridge. Again, we hung around the radio that Christmas Day as news of the tragedy unfolded in a country area maybe 87 miles east of us as the crow flies. We listened in as the Queen broadcast her Christmas message to the Commonwealth ending with personal words of sympathy to mourning relatives and friends. Prince Phillip's schedule was rearranged so he could attend the state funeral for many of the victims.

While this was a sobering time in our little country, I was awakening to the bonds of the British Commonwealth and recalled another tragedy that had impacted me earlier in the year. The blizzards and the hardships of Captain Robert Falcon Scott and all who went with him to the deep frozen mysterious south came into my life when a school trip enabled me to see the movie Scott of the Antarctic, starring John Mills. Robert Jackson (White Continent, Collins Annual 1952) said it this way:

> *Scott, Shackleton, Amundsen—these are names that bring vividly to mind tales of high courage and adventure in the wild snow wastes of Antarctica—the 'white continent'—vast, cruel, fascinating.*

My high school years (1954-57) did little to improve my unsettled, unfocussed attitude. For the first two weeks, under the blazing hot summer sun we wore the itchy, scratchy khaki serge uniforms of the army soldier. Cadet military training was part of life, New Zealand's compulsory military training response in the wake of World War 11. Learning to march, learning to carry a rifle, learning to form lines and learning to say "yes, sir!" did not sit well with me even though I was a heavy reader of the exciting, daring and miraculous events of the war brave. But the retching horror of war and its prisoner camps and the power of politics unveiled a cynicism in me. I was more than upset that part of our training at some point would actually include shooting targets with real bullets from a .303 rifle. It appeared I did not have any options until I discovered over a doughnut with friends at the school tuck shop one morning that perhaps I should join the medical corps. I applied and to prove this was not some wild whim, I emphasized my first aid badge work successes while in Boy Scouts. For the first two weeks at the beginning of my four high school years plus a half a day once a month, I was a medic, and eventually a corporal. We took part in the excitement of field training, ran around in the smoke bombs, dressed the 'wounded' and carried out and the 'bodies'. On parade days we got to run in and pick up the poor fellows who fainted in the summer heat. I was sent to training camp with real soldiers and that is where "yes, sir!" came to mean something very real. I never did get to fire a .303 rifle.

My escapes from the arguing and noisy conflicts in our eight-member family were before and after school delivering milk and newspapers. School holidays found me sweeping floors and moving fixtures in a department store where at busy times like late nights and Christmas I operated the elevator taking busy shoppers to the upper floors. To keep the peace with the rabid sports jocks at school, I found some time for field hockey and in the summer continued my swimming and surf lifesaving patrols. I'm not really sure how my Dad held it all together for us. Coping with six of us kids and a very sick wife, he went from job to job to get even the smallest of pay increases, and, at one time in those years, held down two full-time jobs, moving from a financial clerk in the day to the city gasworks, shovelling coal and coke to and from the fiery retorts till midnight. I didn't spend much time at high school in my second year, finding the beach and good surf a far more enticing space for my imagination between work hours. The third year, I did try a bit harder but failed so miserably in the national examination, I figured I was destined for a life at the end of a shovel. My French and English results kept me slightly above water. That summer I found a new money making pursuit -- a dairy farm, a marvellous place to spend a six-week vacation milking cows, feeding calves, driving a tractor, and haymaking. My fourth year, a repeat of the third, started rather badly with the headmaster, during mass assembly of 1000 boys, summoning me to his office, where he bawled me out, saying that unless I changed my ways and attitude, he'd have my parents remove me from the school. This forced me to take stock and decide that I'd better knuckle down and do better. My parents really had enough to worry about.

I struggled through the year, still working my jobs before and after school. I resat the exam. In the final weeks and not long after my 17th birthday, my English teacher of his own volition, took me to the career counsellor. Together they arranged an interview for me with the assistant editor of the local evening newspaper. I went along with the idea of a career in journalism but could see my mother's tears of protest. Journalism was definitely not on her list of career choices for me. My elder brother had tried it a couple of years before and abandoned that for a teaching career. Mother's concern for me, I think, was I'd fall into the hard drinking pattern of journalists. I had the interview, really thought it would go nowhere and headed out to the farm for a second glorious rural summer. My Sundays off were spent getting to the surf beach and taking part in my club's regular weekend surf lifesaving patrols and training for the national championships.

In mid-January of 1958, I scanned the daily newspaper over breakfast at the farm willing my name to appear in the national examination results. It did not. I could hear my Mother's scream. The examination was the minimum you could leave school with in the hope of a career job. I didn't want to go back to school and told my boss that I'd like to stay on the farm. He thought that would be ok, too, but added that I was nuts. I was a townie and unless I had parents or relatives with money or who owned a farm, I would never make it to owning my 100-acre dairy farm in paradise.

It was a surprise then, when toward the end of January that I got a call from my dad, now enjoying life as a web press operator at the newspaper, that the editor wanted to see me. Big surprise! I had not told anyone about the interview. Even dad did not know why the editor wanted to see me. My employer told me to go for it and drove me to the bus to get into the city for my appointment.

As a circulation boy for several years, I was well known and knew my way around the hallowed halls of the newspaper building. That day though, I was just a nervous 17-year-old schoolboy walking up the stairs one tread at a time, far different from the days I barrelled up the curved stairway four steps at a time to deliver the first copy press proofs to the editor. The interview proved to be extremely brief. The Editor, a very soft spoken man, noted I had excellent passing marks in both English and French, commented on my earlier interview, and asked: "When would you like to start?"

My suggestion of February 2 was a good one, he said, "as we have another young man starting on that day."

My mother was the next challenge. I had decided my future and dreaded telling her I would not be returning to school. I knew she'd be very upset. The family was enjoying its first-ever beach holiday and had rented a camper at the same beach as my surf lifesaving club. I dropped in at the camper and after the welcome my mother got round to the big issue, vehemently declaring: "Well, you will be going back to school!"

"Actually," I mumbled rather quickly, "I-start-with-the-Herald-as-a-reporter-on-Monday."

It was just as well I was standing in the doorway. I am sure mother's re-action was heard down on the beach. She hollered at my dad to "do something." As the air cleared and things calmed down, my mother remained adamant that I'd return to school. I was the opposite. Dad quietly nodded things would be ok and suggested it might be best if I headed down to surf practice.

With the last bale in and the neighbour's hay stacked, I finished at the farm and went shopping with some of my earnings to outfit myself for the office life, a couple of white shirts, trousers, jacket and tie, socks and shoes.

On February 2, 1958, I grabbed my green three-speed and proudly rode in to report for work at the 106-year-old *Taranaki Herald*, the oldest daily in the country. My first assignments were to check the obituaries for followup stories and learn how to draw the weather map from the coordinates deliv-ered mid-morning by Post Office telegram. My co-starter that day was Lew Pryme, a rock 'n roll singer in the making.

Exactly two years later, I was hammering out a story at my desk behind the door when the Chief Reporter introduced a new cadet photographer to the room. I turned. I mumbled something, blushed brilliantly and went right back to work. The new photographer was the one person I had always wanted to meet and never could! My venerable stone and ivy high school featured a low rock wall alongside the Memorial Gate. Before the daily morning as-sembly, a group of us would sit on the wall to watch the high school girls ride past on their way to school. And that is when I first saw Lois, head down, cycling furiously past, ignoring the yahoos on the wall. Most fine days, until the wall was declared off limits due to unseemly behaviour, I'd watch her go by, declaring on one occasion that she was the girl I would marry. "Hah," shot John who lived only couple of blocks from her and knew her, "She wouldn't marry a pumpkin like you!"

I got over my initial embarrassment and life returned to normal in the collegial atmosphere of the newsroom. After a few weeks the Chief Report-er delegated me to teach Lois how to ride the office motor scooter. Though among the unreachable at school and church dances, as work colleagues we got to know each other. Then came the weekend my current girlfriend didn't

Lois at play in the garden.

Hilary and Rachel enjoy camera time.

Bridget finds her walking legs.

want to go to the Saturday night dance. My school pal Roger Quail was in town from his out-of-town farming pursuits, so we went anyway. As I walked in the door of the Star Gymnasium there she stood. Lois greeted me with a shining smile and I asked her to dance; we stayed right on dancing. In the coming weeks, each of us had breakups to deal with but we already knew we'd stick together. Our quiet, definitely-out-of-the-office romance blossomed. We were still a ways from the altar, with a lot of planning and saving to accomplish. Parental approval was needed, too, as we were both under 21. Some 10 months later, in April, 1961, we exchanged rings and vows in an historic stone church, enjoyed a marvelous two-week honeymoon and embarked on the thrill of setting up home in a dilapidated rented house. A year later we upgraded to a swankier apartment suite in a red brick mansion and furnished it as best we could for the arrival in June, 1962 of Hilary, our little red-haired daughter. Her birth sent us over the moon.

Outwardly, the early 60s were good years to be living in New Zealand, a country with extremely low unemployment amongst the 2.5 million people. Our islands had come a long way from the big depression and world war years. Social politics gave its citizens a modest standard of living. The green grass and high country yielded butter, cheese, meat and wool for foreign markets. By the spring of 1964, Lois and I, through savings, sweat equity and government incentives, moved into our own three-bedroom home on a 1/5 acre lot. Within a couple of weeks of moving and a day after my grandfather died, Lois went into labour and Rachel arrived. We were a happy family. My job was going well with port development, oil and gas exploration and new industrial development. We had a good and vibrant privately owned afternoon daily newspaper with its heart in the community. Most days, we outclassed the opposition morning daily. But in the mid-60s rumbles started on the country's economic front. The traditional United Kingdom market for New Zealand's butter, cheese, meat and wool was on a slippery slope with the continued talks of the UK joining the European common market. Pirate radio was challenging the government broadcasting monopoly, the sterling gave way to dollars and cents, and the bars moved from 6pm closing to 10pm. A dramatic 30 per cent drop in wool prices socked the country into an economic depression in early 1968 and the country staggered into unemployment and inflation.

To make matters worse, Cyclone Giselle ripped in from the tropics and wreaked havoc from the top of the North Island and through the South Island before dying out somewhere in southern latitudes. Unfortunately its 200 miles an hour winds collided with a southerly storm over Wellington, forcing the inter-island ferry Wahine to smash on to a reef as she entered Wellington Harbour in the early morning after an overnight voyage from Lyttelton. The vicious storm claimed the lives of 52 of the 734 people on board. A month later a major earthquake at Inangahua on the South Island's West Coast left three people dead.

Dwelling on the brute unexpected force of flood, storm, sea and 'quake and the human balance within the unpredictable borderlands, I recalled the times I had sworn and argued with the pelting rain and freezing wind as I cycled home from delivering papers, a newspaper shoved up my sweater to stem the frightful cold. Yet, when I arrived home, drenched, there was a certain feeling of victory, of having won out against what nature had thrown at me that afternoon. I loved watching the storms on the harbour and walking the beaches in the sting of the salt-laden spray. And there was always the sense of real adventure climbing through the bush on the mountain in the clinging, damp cloud.

My curiosity enabled me to progress rapidly at the newspaper and I soon graduated to handling wire news as a young sub-editor. It was very apparent that no matter where you were in the world, weather was a major newsmaker. The dispatches of Sir Edmund Hillary and Sir Vivian Fuchs and their exploits completing the TransAntarctic Expedition in 1958 were in my hands each day before the newspaper hit the stands. Weather stories were mostly always front page, flooding, gales, downpours, earthquakes, and record snowfalls on the mountain. The newspaper was an exciting place to be and I loved the gathering, placement and daily scramble to produce a new edition on time each afternoon. I worked to a daily deadline, procrastinated until the adrenaline took control and bounced between the highs and lows.

World events were changing New Zealand's economy right down to the profitability of evening newspapers. I found myself up against increasing conflicts simply because I could not accept the 1962 merger of the privately owned newspaper with the opposition morning daily. The "family" I'd been part of for 13 years, from the ground floor up, was changing. The first couple of years had been ok, but with deteriorating revenues, consolidation was necessary. The changes bothered me.

One bounce saw me quit the newspaper and take a position as one of the original nine regional journalists with New Zealand Broadcasting. I stuck at that for a couple of years but headed back to the newspaper so I could "write the whole story" instead of a paragraph or headline for radio news bulletins.

Each year the New Zealand government advertised for candidates to man the sub Antarctic weather station at Campbell Island, almost 500 miles south of New Zealand. Around the same time, there was a call for candidates for the ongoing New Zealand Antarctic Research Programme (NZARP). I read them and tossed them aside. I was now a family man. My Antarctic dreams never quite faded and when the opportunity arose in 1968, Lois and I saw the personal and career benefits that could accrue from such an undertaking. Add to that, a close colleague took off for a six-month special assignment with the New Zealand Press Association (NZPA) covering the Vietnam War. I had that marooned-on-a-desert isle feeling. I enjoyed what I did to put bread on the table, enjoyed getting up every day and enjoyed working with great friends in a pleasing environment. I valued the independence that enabled me to set my own assignments most days. I had full responsibility to keep the burgeoning oil and gas industry on the pages of our newspaper. I stayed out in front, garnering the news ahead of the competition. I spent several days on an offshore seismic exploration vessel and worked a supply boat on a couple of days off from work. My editor put me on a joint New Zealand-Australian navy exercise in the boisterous Tasman Sea. I put to sea on day trips with government fisheries technicians probing currents and temperatures and fish migratory patterns.

But happy? Deep down I knew there had to be something more. As my mother had feared I was drinking far too much and failed to recognize or heed the advice of my father-in-law: that one day I wouldn't be able to stop. My habits meant I was riding very close to the boundary all the time. Little by little I was beginning to begrudge life in New Plymouth. I was in a life-style that was going in the wrong direction and I had no idea what to do about it. I was itching to do something else or be somewhere else. Into this 1967 mix came the wonder of our third daughter, Bridget. Like Rachel, she was a little blondie who allowed us to see above the intruding background noise of life. My anchor was Lois and our daughters, yet in spite of that, the spontaneous after-work drinking parties with my colleagues continued to be major draws. Perhaps, I was becoming two people: a happy family man and a hard-drinking journo. I allowed stuff to get in the way of Lois and I carving our initials on the foundation block of the future we had promised each other. I knew the track I was on was destructive, to myself, to Lois, and to people around me. Something had to give. Change was paramount.

The explorer who goes to Antarctic must take everything with him. In the Antarctic he finds nothing to aid him; no food, unless he fishes in the icy seas around the continent; no firewood; no stones for shelter...He is friendless in a sullen, antagonistic continent. So he must take his own tents and sleeping bags, his own food—easily carried, nourishing food like chocolate, dehydrated vegetables, rice, cereals, powdered milk, dried apples beans and turnip tops, his own sledges so he can move with speed and ease, his own fuel, candles for light and thermos flasks to keep his drinks hot."

...Robert Jackson, The White Continent.

Antarctica beckoned. I had Lois' full support and agreement.

US Navy Super Constellation Pegasus at the McMurdo Williams Field airbase.

2.
ON MY WAY

It was my birthday, Sunday, October 13, 1968.

I was headed south from Christchurch, New Zealand, aboard Pegasus, a US Navy Super Constellation aircraft, the largest plane I had ever flown in. Military style, we faced rearwards. My legs were cramped from sitting hump-kneed over my duffle bag wedged between the seat rows, uncomfortably stuffy and hot. The four propeller engines droned on hour after hour to deeper latitudes.

The excitement of the sheer enormity of this adventure into perpetual cold, ice and snow at the bottom of the world eased the pain of birthday over-celebration the night before. There it was, McMurdo Sound, the South Pacific entry point to the vast continent of Antarctica, where I'd begin my five-month "summer" assignment as information officer/photographer with the New Zealand Antarctic Research Programme.

We'd already partially done this 2400-mile flight the day before. Same thing, same drone, same cramped seating, same excited anticipation of adrenaline-pumping adventure. But at the point of safe return we'd turned back to Christchurch because of deteriorating weather at Williams Field, the US McMurdo Station airbase. The fuel range of the Super Constellation was for a one-way trip. We'd have to wait another 24 hours for an improved weather window.

It was a different way to travel – five aspiring New Zealand polar explorers stacked into a planeload of American servicemen, civilians and supplies. We were the fresh faces going in to replace the winter-over beards and to deliver mail, news and fresh food. About one-third of our way through the 13 hour journey, we dissected military rations for a lunch-cum-snack. Discovering what soldiers found in the khaki tins and packages provided moments of joking relief with my new friends in the sardine row.

It was springtime on the ice. And that meant the United States and New Zealand Antarctic interests had begun the massive annual effort to re-people, restock, replace, renew, repair and prepare for the next winter of dark isolation.

What would I find during my tenure, and what would I become after a summer in that desolate, windswept, inhospitable mass of ice at the bottom of the world?

A challenge by Howard Marriott, my neighbor at home in New Plymouth, New Zealand, precipitated my journey some 10 months or so earlier. In the middle of summer, we were landscaping our new homes when he suggested we sign up for a season on the ice. Not a bad thought in the heat of the summer sun.

Next thing I knew Howard and his wife Gin walked through our door, demanded a cup of tea, and the four of us sat around the dining room table. They had the advertisement with them setting the scene for a lively discourse on the pros and cons of abandoning life as we knew it to take a career step onto the Antarctic ice.

I admired Howard as a risk taker. In his late teens, he'd travelled to New Zealand with his family under the assisted immigration scheme and once in the country followed up on a contact he'd made in England and went to work in the fledgling table chicken industry. He'd left his girlfriend behind but she followed as soon as he had saved enough money for a full fare passage. They married, settled on the chicken farm and all was well till allergies got the better of our curly-headed friend. He took a diesel mechanic's course, got his ticket, built a house next door and we became lifetime friends.

Howard and Gin planned to sell their house. Howard would head south, and Gin would spend most of the year with their two wee children in England visiting family.

"Well, what do you think?" Gin said, laughing. "You going to do it?" Because of my whining that we were too poor to buy me trousers, she built on her encouragement by promising to buy me a new pair.

"Yeah," I replied. "In your dreams. We've got houses and families. You've got two kids and we've got three and our little one is only a few months old. How can we leave home for months and months?"

"Why not?" chimed Lois.

The seed found soil. It germinated, demanded fertilizer. Howard led the charge, and brought home the application papers. Together we filed our dreams and waited. We checked off the weeks. Nothing happened. We were now well beyond the closing date.

"Perhaps its because we are married with children," I rationalized.

"Have patience," Howard said. "We're talking government!"

My letter arrived, nothing in Howard's mailbox. I was called for an interview at Antarctic Headquarters, a 250-mile railcar ride to the country capital of Wellington. The New Zealand Antarctic Division was part of the Department of Scientific and Industrial Research (DSIR). A country boy within the halls of bureaucracy, I behaved myself, gave blunt short answers to the panel's questions, sweated from nervous excitement and had to think hard about a question from the leader-elect, Robin Foubister. What habits or traits would bother me about other guys I was living with at close quarters for a long stretch of time?

"Maybe someone who eats with his mouth open or slurps his soup," I countered, after a long pause, recalling my Mother's admonitions (and sore ears) to me about sloppy eating habits.

"What would you do about it?" he asked.

"I'm not sure," I said, trying desperately to think of a good answer. "Probably nothing. I could choose another table, but not really. I'm sure I'll do things that bug people."

It was a long train ride home. I went back over the interview, and although my life's work involved interviewing people for newspaper stories, I was sure I had not been an exceptional interviewee! I was thoroughly intimidated by the whole process with two, or was it three, people seated behind a table and me sitting in a lone chair in front of them. Foubister led the questioning backed up by Antarctic Division boss Bob Thomson, an experienced Antarctic man with Australian and New Zealand expeditions. His early days had been spent growing up in a small town a few miles north of New Plymouth. As the train clackety-clacked over the rails, I kept going over some of the questions. What was the answer I had given when asked about my drinking habits? And that half-truth about owning a camera...what was that all about? I knew the model we owned would not be up to the job and we'd have to buy another. The weeks passed and I interpreted the silence as the government's way of saying "no thanks".

Again, when I least expected it, another letter arrived and I learned I'd been shortlisted from a list of 90 applicants. Sadly, there was no such letter for Howard. We would not travel together and share in the adventure. (Good news came later, though: he was selected as Base Engineer for the following year and would winter-over in Antarctica).

Even though my application was almost successful, there were still a few hoops to go through before I knew for sure that I would spend a long summer in a remote place away from Lois and our three daughters now 6, 4 and 16 months. On the official side I needed the prerequisite medical clearance. On the personal side was the reduced income. We were already struggling with the bills on my journalist's pay. An unsuccessful career move to another city a couple of years earlier had left us with heavy car repair payments and arrears with our mortgage as the renters had not paid up. Staying home, consolidating and keeping our expenses low made perfect sense. Lois did not drive and really didn't want to anyway. How would she get around? A new camera was necessary. While travel to and from training and orientation would be met by the government, we'd have extra personal expenses. On the plus side most of my clothing and all living expenses would be provided. That is how we rationalized the income reduction. Even though they did not really acknowledge the purpose and the opportunity, our parents and families were very supportive. Together with the promises from friends and neighbours, Lois and the girls would be looked after. One promised to continue her driving lessons.

The Antarctic Division's method of announcing successful appointments proved extremely embarrassing. A government-issued news release on a Friday evening and a call from the local reporter for radio news was the first indication that I'd made the final selection. The announcement of the 1968-69 team was national and local news. The morning newspapers carried the story.

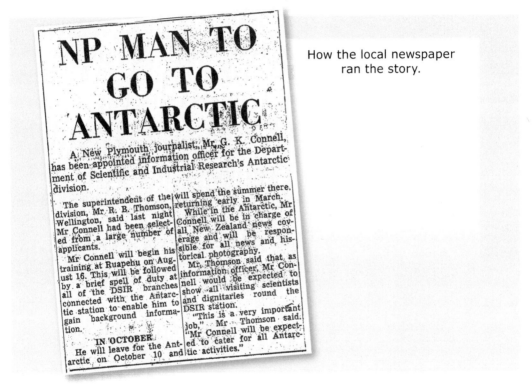

How the local newspaper ran the story.

NP MAN TO GO TO ANTARCTIC

A New Plymouth journalist, Mr G. K. Connell, has been appointed information officer for the Department of Scientific and Industrial Research's Antarctic division.

The superintendent of the division, Mr R. B. Thomson, said last night Mr Connell had been selected from a large number of applicants.

Mr Connell will begin his training at Ruapehu on August 16. This will be followed by a brief spell of duty at all of the DSIR branches connected with the Antarctic station to enable him to gain background information.

IN OCTOBER

He will leave for the Antarctic on October 10 and

will spend the summer there, returning early in March.

While in the Antarctic, Mr Connell will be in charge of all New Zealand news coverage and will be responsible for all news and historical photography.

Mr Thomson said that, as information officer, Mr Connell would be expected to show all visiting scientists and dignitaries round the DSIR station.

"This is a very important job," Mr Thomson said. "Mr Connell will be expected to cater for all Antarctic activities."

The trouble for me was that I was employed as a senior journalist with the evening newspaper in our home province. I told the radio news reporter that as far as I was concerned, my selection could not be validated until after a successful stringent, military-level medical examination. I tried to water down the morning newspaper reporter's questions with the same line. Both reporters were ecstatic: they had both scooped my own newspaper. My newspaper would be a distant third with the news. Usually I bounded up the stairs two or three at a time, but this Saturday it was a very long, slow walk. Whose face would I see first? The morning newspaper would be opened and on everyone's desks. I'd already seen the story: 'NP Man To Go To Antarctic'.

"Shit, Connell! You don't do things by halves!", came a faceless voice from the Chief Reporter's room. Red-faced, I slipped into the Editor's room and heaved with relief at the smile and congratulatory handshake. By the end of the day I was glad it was Saturday. I was on a high all day taking phone calls from friends and my business contacts. I don't recall any naysayers. Many asked after Lois, our daughters, and post Antarctic life. I went home exhausted.

The official news release detailed that I would be responsible for news coverage, and all news and historical photography. I would also take care of all visiting guests of the New Zealand Government, including scientists and assorted dignitaries showing them around the base and outlining New Zealand's scientific endeavours and achievements.

Some weeks later, I was called to training camp. In return for a week off without pay, I promised Editor Rash Avery a full-page exclusive story and pictures on the weeklong mid-August training camp, two months ahead of departure. The Taranaki Herald ran that feature on August 31, 1968. It outlined all the physical and mental preparations in readiness for a season on the ice, isolated, fully dependent on each other and our own resources.

I met my fellow explorers at Waiouru, a New Zealand army camp in the centre of the North Island. The programme kicked off with the expected orientation to the then Department of Scientific and Industrial Research Antarctic Division, what we would find at New Zealand's Scott Base and an appreciation for New Zealand's scientific activities in Antarctica. With fire being one of the great perils of polar life, we had a very thorough grounding in safety and prevention as well as very practical hands-on instruction on how to tackle a blaze and put it out. The week's survival training was interspersed with lectures on geology, fauna, personal health and hygiene and DSIR administration (how we'd get paid, for instance).

For four days, we took World War II vintage army transport to the nearby rugged, snow covered and unpopulated slopes of Mt Ruapehu, the highest peak in New Zealand's North Island. We were realistically exposed to working, traveling and surviving in snow and ice – pitching polar tents, camping, field radio operation, field rescue and first aid. It was thorough. It was great. We had fresh snow, wind and cool temperatures but the one thing we discovered later, the single thing that could not be included in our orientation, was bone chilling Antarctic cold. Our ski instructor kindly pointed out that he did not intend to turn elephants into champions but to teach us how to use skis as a basic form of transport. We even had to hurl ourselves down a slope to practice "self arrest" using only an ice axe to stop sliding to an icy end. Intentionally hurling myself into nothing, landing and rolling onto an ice axe to stop the slide was certainly challenging. I was thankful to have a couple of funny guys in the group who made light of my awkwardness as well as everyone else's attempts. With ribald humour, they "scored" each jump so by the second or third attempt we got closer to a 10. We were raw talent!

Realistic training on New Zealand's Mt Ruapehu. Finding out about polar tent erection in snowy conditions are, from left, Leader Robin Foubister, Hugh Clarke, Doug Spence (I think) and Alister (Taffy) Ayres.

Early friendships were evident as we worked together under the trying winter conditions of driving snow and freezing rain. All together 36 Antarctic veterans and novices took part in the camp – 24 from the DSIR including the core summer and winter team, 10 from the Victoria, Canterbury and

Otago university research programs and one representative from the Canterbury Museum.

The 1968-69 New Zealand Antarctic Programme (NZARP) group was ready to go and comprised:

From Auckland: Keith Mandeno (technician, winter), Nigel Millar (technician, winter), Wayne Maguiness (fitter/mechanic, winter), Geoff Gill (cook, winter), Alister Ayres (assistant surveyor, summer), David Blackbourn (Post Office technician, winter).

From Thames: Brian Hool (Postmaster, winter).

From Taupo: Chris Rickards (base electrician, winter).

From New Plymouth: Graeme Connell (information officer/photographer, summer).

From Wellington: Bob Hancock (radio operator, summer).

From Christchurch: Robin Foubister (leader, winter), Peter Lennard (technician-in-charge, winter), Bruce Brookes (assistant maintenance officer, summer), Doug Spence (storekeeper, summer).

From Dunedin: John Newman (carpenter, summer), Charlie Hughes (surveyor, summer).

From Fairlie: Allan Guard (base engineer, winter), Derek Cordes (field assistant, summer).

From Invercargill: Hugh Clarke (assistant maintenance officer, summer).

From Wanaka: Noel Wilson (field assistant, winter).

From Timaru: Bill Lucy (deputy leader and leader, Vanda Station), Ron Craig, Warren Johns, Simon Cutfield.

It was an introspective group that scattered to all parts of New Zealand at the end of that brilliant week, knowing that our next beer together would be "down there". Meanwhile, from what I had learned and talked about, I felt better equipped emotionally to get my personal and family life in order during the six weeks before departure. This week introduced change. I had busted out of the vortex I was in and had a new confidence that Lois and I could bring our debt under control and get on with the business of life.

The full weight of NZARP that summer was the establishment of a new year-round scientific station at Lake Vanda, about 80 miles northwest of Scott Base. The unique Vanda Station would be New Zealand's first mainland station in Antarctica and would be the smallest to be manned through the long winter months since the heroic (pre-1925) exploration era. Other than that our team would maintain existing ongoing programmes, and provide support for the continuing university projects. All government-sponsored activity had been cancelled in favour of Vanda. Underneath this of course would be caring for New Zealand's international presence. The camp had introduced me to modern Antarctic living and survival. Our own appreciation of the history of the place would fill in the gaps. As good keen Kiwis (New Zealanders) we knew the stories of the early explorers who staged out of New Zealand. You might say it was part of our cultural heritage. In our eagerness, enthusiasm and naiveté we too hoped to become known as intrepid Old Antarctic Explorers – at the very least to our families and friends.

Waiting for our ride to the ice at Christchurch Airport, October 1968, from left: Bob Hancock, John Newman, Noel Wilson, Doug Spence.

At the end of September I finished my job at the newspaper to spend a couple of weeks in Wellington with my new employers and meet the people I'd be working with long distance over the next 21 weeks.

The training camp also showed me just how much I did not know about cold weather photography and how much I did not know about the workings of my new camera, a Canon FT single lens reflex with a 135mm Soligor telephoto lens. Just prior to the camp, Lois and I had stretched out on a financial limb to buy this state of the art camera thanks to the efforts of creative financing put forward by the camera store guys and friends Peter Hamling and Gordon Beckett. The camera also sported through-the-lens metering.

My preparation complete, I now shared a row of military plane seats with four of the blokes I would spend every waking moment with over the next five months – carpenter John (Chippy) Newman, storeman Doug Spence, field assistant Noel Wilson and radio operator Bob Hancock. One group had already gone down a few days ahead of us. The rest of our team would follow at intervals over the next couple of weeks.

I might have embarked on the adventure of a lifetime but what about Lois at home with three young daughters, a Basset Hound, a moody car, and her blizzard of bills, the bank and a mortgage? I knew I had to keep her up to speed as to what I was doing to lift her spirits up and support our own private journey. Trouble with me was I had never been a letter writer. Now I had to make the time and effort just so that the whole adventure would be a shared memory for all. She would be bombarded everywhere she went in our hometown with "How's Graeme?" She had to have answers. On top of that, we had no idea where this journey would lead; we had no idea where I would find work when it was over, least of all where we would live and grow. One of the final things I did before leaving home was to call on a real estate friend and tell him we'd be selling the house but not to list it. If a buyer came in we'd be interested and deal with it at the time. In other words, we were keeping our options open.

I wrote her during our 24-hour wait:

October 10, 1968

Hi Hon:

Well, this will be the last letter I write to you from New Zealand. And I got your last just before leaving Wellington. It was good to get. I think it would be best if you addressed your mail from now on to me c/- Scott Base, Antarctica, c/- CPO, Christchurch. After all, I will be down there on Sunday, for lunch. Latest information is that the plane leaves Christchurch at two minutes after midnight. What a start for my birthday!

* * *

We are in a waiting game I had to dash off in the middle of things yesterday and was on the go all day visiting here and there. I called on John Whalan at the DSIR photographic unit and learned quite a bit from him. John goes down to Scott Base sometimes to take official pictures. He's only in and out though and has never stayed long –like I will be. Caught up with my old friend Lieutenant Jerry Power to chat about the Navy side of things and later we met with Dominion and Evening Post editorial staff fellows over a beer. They gave me heaps of encouragement. This morning I went and spoke with a former polar photographer at Kodak and got a few more tips and what conditions are like down there. No need for any filters for colour apart from an ultra violet that he says I should have on the camera all the time anyway. For black and white, a yellow would be useful and possibly an orange. It is worth two stops. I'll go shopping in Christchurch as I should have a bag but I don't know what funds I will have left.

I might get a ride in the Boeing 737 tomorrow. It is not on the run officially yet but it has been operating.

Bye Bye My Sweet

G

* * *

In Christchurch, I settled in for a couple of days at the Wigram Air Force base, very handy to Christchurch Airport, the hub of US and New Zealand Antarctic operations. There I telephoned Lois and we chatted about nothing like a couple of teenagers trying to organize a date! Neither of us really wanted to signoff. Our parting really did sound as though I was going to the ends of the earth. We enjoyed many phone calls through that weekend, and I gave her a great surprise on the Sunday when she fully expected me to be sitting next to a penguin at Scott Base. I was still at Wigram. It was a short call just to let her know I was still in New Zealand. With long distance charges in mind we kept this call short and avoided slipping into talk of our time apart.

* * *

We knew we were getting close to the end of our 13-hour flight when the scratchy public address system advised us to change from civvies into our Antarctic gear if we didn't want our butts to freeze. We sweated in the down clothing for almost an hour, but 30 minutes before our landing at McMurdo's Williams Field the temperature inside Pegasus took on a definite chill. We craned at the tiny porthole windows for a glimpse of ... well anything really, but most of all Scott Base. It was ice to the left and ice to right and ice below, all of it white.

The flight deck reported that visibility had deteriorated and ground level conditions were close to a whiteout. The point of safe return now several hours behind us, the PA crackled and with the laconic advisory to brace for landing; "There might be a slight bump." I braced, trying to find some comfort in the fact that the flight crew and the plane had done this before. But then again, my brain argued, this was one of the early flights of the season and it might be a rookie crew.

All I knew about Williams Field was that it was smoothed out of the ice and groomed to allow non-ski planes to land. The ice might be up to 20 feet thick. Under that was water, like ocean water! Not much use looking out the porthole. I was a few seats away and wouldn't see much anyway. "Brace," the PA announced again. I'd flown in small planes and DC-3 passenger aircraft landing on grass runways and paved runways. But this was ice. And this was a big aircraft, the kind Trans World Airlines and Pan Am flew around the world.

A tense hush settled over the passengers as the big plane seesawed from side to side, up and down, decreasing altitude for a landing. I didn't like this swinging around. I hoped I would not need the thoughtfully placed paper bag. The porthole windows just showed cotton wool. The near-whiteout conditions meant that the ground and sky were the same colour, white. Our flight crew and instruments had to estimate the distance from wheels to ice runway; we dropped hard the final 12 inches or so, a major jolt, a small bounce, and a roll to stop at the end of the runway. An audible sigh of relief settled over the passengers who erupted into boisterous cheering for the crew as Pegasus turned and taxied to our Antarctic terminal.

The scramble to deplane began. NZARP Leader Robin Foubister met the five of us as we clambered out, struggling with our gear and squinting in the bright whiteness of our new surroundings. I confessed bravely to not yet feel the cold, as it was a balmy minus 20°C with a wind. Though this was the coldest temperature I had ever experienced in my young life it didn't take my breath away. What did

Magnificent Mt Erebus, at 12,448 feet, it is the world's southernmost active volcano. This view greeted passengers in 1968-69 as they deplaned at Williams Field.

take my breath away was the white, we were wrapped in white.

The tiny window view had given me a mere picture-frame glimpse of the immensity of frozen McMurdo Sound. It was very difficult to appreciate that what we could see of the Ross Sea in every direction was a layering of feet thick ice over the deep blue. Would I ever get to see water? No trees or natural landmarks in this barren, frozen land. Flanking the runway, the bright orange Quonset huts and vehicles of the US Navy's Williams Field air base appeared almost out of place in this stark environment.

I just could not get over it. Even with the limited visibility, I gazed wide-eyed over my new surroundings, seeing the ice and snow fold and roll over the landscape. It was riveting for a fellow whose view of ice and snow had been limited to a couple of volcanoes in New Zealand's North Island. How would I ever describe this to Lois and the girls? Really, what colour is white? Is there light white, dark white or even bright white? After all, we define other colours in shades, like we have bright green, dark green or light green.

With our gear loaded into a rubber-tired long wheelbase yellow Land Rover, Robin chauffeured us at 30 mph across an almost "super highway" of ice to the "Granny Smith" apple green huts of Scott Base that stood stark against Pram Point's white backdrop. The New Zealand base lived up to the planners' expectations I'd read about during my indoctrination in Wellington. When the base was established in the late 50s, it had followed the standard highly-visible yellow and orange colouring but two years before our arrival the base was changed to green to look like the New Zealand countryside.

Questions flew from our little group of polar neophytes -- what's this, what's that, how cold? Then we went silent, just to hear the thrumming of the tires on the ice road, gazing out the windows as brown skua gulls circled above the ice in search of supper. I was stunned and overwhelmed at the realization I had actually put two feet on Antarctic ice, a place where only my imagination had been.

We parked at the old hangar and lugged our baggage over the snow to the Base complex. Non-stop chattering broke out. Look at this! What's that? We wanted to know about the massive chunks of ice strewn haphazardly in front of the base like wayward sugar lumps that had leapt from their bowl and grown into giants. "Pressure ridges or what we call the tide crack," our leader explained. "...the ridges are formed by the thick sea ice meeting the land and buckling under the tremendous tidal pressure. It's a dangerous place and generally out of bounds. You will hear more about that later at our safety and orientation meeting."

Scott Base was established in 1957-58 for a five-man International Geophysical Year (IGY) scientific team and as the base of operations for Sir Edmund Hillary's activities for the successful Commonwealth TransAntarctic Expedition (TAE) of Britain's Sir Vivian Fuchs. The science data (including auroral, ionospheric and seismic) provided a link in the research activities at the US Pole Station and the joint NZ-US station at Cape Hallett at the continent's northwestern entrance to McMurdo Sound.

The buildings were all prefabricated in New Zealand to ensure speedy construction. Prebuilt in New Zealand, each component was numbered as they were dismantled for shipment. Erection at Pram Point involved site preparations and securing of fastenings. The base was extended to 10 huts in subsequent seasons and New Zealand continued to operate Scott Base to

New Zealand's Scott Base with the tide crack and pressure ridges in front.
Plenty of Weddell seals lying around too.

maintain the continuity of scientific data.

After the hangar, we paraded through the dimly lit garage and emerged into the start of the corrugated iron arches of the interconnecting 150 meter (500 feet) covered way to all 10 squat, flat-roofed huts in the base complex. From there it was through the workshops, then the ablution block, past the hospital, past the main sleeping hut, and into the Post Office, Radio room,

Reception and what was to be my desk. I dumped my belongings there and followed the group to the mess hut (officially A Hut). Further along the walkway branched off to the cool store and food store with the laboratory at the very end of the line. Outside and some distance from the laboratory were the two free standing seismic huts, the auroral hut and two magnetometer equipment huts. I wandered around the complex to get my bearings, tracking in and out of huts and along the duckboards of the covered way. I located the bunkrooms and workspaces once again and figured out the double entry cold porches (similar to mud rooms) to each hut that branched off the "tunnel". A prime reason for the base being split into various huts was in case of fire. I noted the pile of correspondence on the desk casually labeled P-R-O

Scott Base's covered way, the backbone linking all 10 huts.

(Public Relations Officer). The letters were mostly from school kids in New Zealand looking for answers for their class projects. Correspondence to deal with....life on the ice had begun.

Dinner that first night was a great three-courser that would stand up against any restaurant I knew. Starvation would not be one of my worries. As soon as I could, I phoned home. Lois and I had been apart for just one week. We now had to learn how to converse over a radiotelephone link ("hello sweetheart...love you... over") and preserve every precious minute. Another 21 weeks before we'd see each other again. I knew the girls were missing their dad. Hilary had an understanding that I might be gone for a while and that I was in a special place while Rachel figured I'd be home tomorrow. Bridget had not yet learned to talk. I felt really alone now as I wandered out of the communication hut and back up the covered way to the mess hut, officially A Hut. Self-doubt surged through my fatigued, overwhelmed brain as I recalled the mountains Lois would have to climb and the glaciers she would have to traverse. The next few months would be a challenging learning time for each of us.

My gloom dissipated in A hut where I was surrounded by the bearded ones wanting to go home and the clean, fresh-faced neophytes eager to begin a new chapter. I listened as the expedition shrink Tony Taylor gave a talk on the art or otherwise of tattooing (any relevance to our expedition escaped me, but he was funny and very entertaining) and then we oohed and ahhed over a slide show of hiking in Peru. The beer flowed and I got a quick introduction to Scott Base style social life and entertainment.

With our room assignments, I received a nice room at the very end of the 1000 square foot main sleeping hut divided into 11 two-person curtained bunkrooms. Two in a room during the summer reducing to one per room in

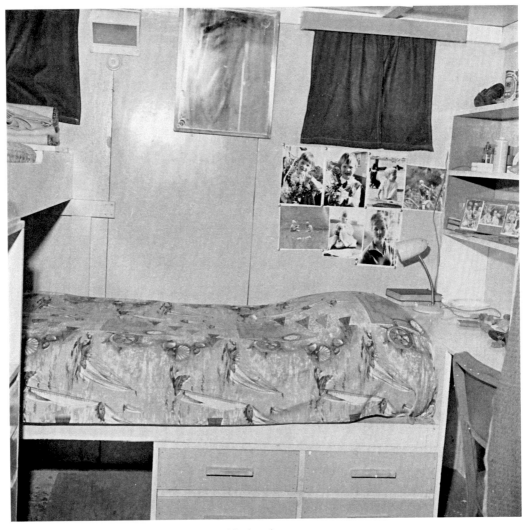

My bunkroom

the winter. My delight at finding I did not have a roommate was short lived as I learned the spare bunk would be for guests overnighting!

I chose the top bunk in the absence of any opposition and unpacked my gear. One advantage about the top bunk was that I was close to the escape hatch – a door about 30 inches square with the same great chunky silver handle you'd find on the door of a walk-in freezer. The hatch was angled on all edges to plug snugly into the hole in the foot thick outer wall. Later I opted for the lower berth as it was better suited to my erratic sleep schedule.

October is a month of diminishing sunsets in Antarctica and there was certainly a gorgeous one this first night. I went outside for "fresh air" and shot a few pictures to send out on the next day's plane. I got a picture of our carpenter horsing around "sweeping" the front door step of fresh blown snow against the backdrop of orange and pink skies, dark clouds backlit by the fast setting sun. Chippy and I stood there in the absolute silence interrupted every now and by the crack of the buckling sea ice less than 100 meters from our front door.

In *The Worst Journey In The World*, Apsley Cherry-Garrard wrote "*.... as you stand in the still cold air you may sometimes hear the silence broken by the sharp reports as the cold contracts or its own weight splits it. Nature is tearing up that ice as human beings tear paper.*'

The corridor in the main sleeping hut.

Back inside the complex, I sauntered quietly to my room attempting to decompress from the over stimulation and excitement of a very emotional day. Almost a year in the making, what would tomorrow hold as I changed gear and wrestled myself into a new job?

I pinned a fresh picture of Lois and our girls to the green wall beside my bed, gazed at it as I jumped into my PJs and slipped into bed and with a cheery "g'night" to them, I flicked out the light. I lay still in the dark quiet of the hut. Fingering the small gold cross Lois had put around my neck when we were last together, I drifted into another world.

It was darker and quieter than an abandoned coalmine when I awoke, disoriented, unsure of where I was. Was it day, was it night? No one to snuggle with and the bed was narrow. Slowly my noggin eased from park to first gear. I lay still and quiet as excitement bubbled through my every nerve. I couldn't remember where the light switch was but I was standing on floor anyway. I dressed in my Antarctic issue clothes of the day before and headed down the black hallway, alive with snoring. I had no idea of the time but being a normal morning man, I figured it was get-up time of around 6 am. I fumbled noisily through the cold porch and into the lighted covered way. Which direction was the ablution block? Left or right? Up or down?

A corner of the ablutions hut, complete with an agitator wringer washing machine!

I chose left and padded slightly downhill along the duckboards to end up in the brightly lit hut that housed toilets, washbasins, a washing machine and a shower. My internal clock still functioned with accuracy. It was just after 6 a.m.

A hut or mess hut where we ate, entertained, socialized, relaxed and played.

"Old habits die hard," I confided to Doug at the row of hand basins, his face under soap, "This is the time I always wake up." Doug was stripped to the waist and confessed to using a basin full of water for a modest body wash. Water was a precious commodity as we were to discover later in the day. It had to be conserved and used at an absolute minimum.

Before I could worry about getting the sleep out of my eyes I had to pee. The base toilet system had been explained the night before but now I was able to follow the full impact of the instructions. The rule was simple: pee first, squat second. O.K., that mean't peeing into a small trough that gravity-trickled through a heated pipe system to the outside. The trough was located at one end of the hut in a small cubicle housing the honey buckets – best de-scribed as an indoor toilet with an outhouse or long drop design! The honey bucket was merely a 44-gallon drum cut in half with handles welded to the

P-R-O's desk in a corner of the Post Office/Communications hut.

side. It resided in a cabinet under a traditional toilet seat. Our orientation the night before was blunt and to the point: solid's only in the honey bucket, and liquids in the trough. Liquids and solids were not to share the same repository.

My background and that of the other newbies was the main focus of discussion over breakfast – oatmeal, eggs, bacon, toast, OJ, coffee, the works.

Between mouthfuls the fresh-faced ones new on base wanted to know how "things run around here" while they, the bearded ones, wanted to know "what's happening out there." I did something I had not had a chance to do the night before and checked out this 500 square foot art deco space with its hardwearing laminate and vinyl furniture. I slowly gazed around the room – the notice board as you came through the cold porch, the entry to the galley, aka kitchen, the crockery and tableware cupboards and counter beside a stove, a dart board, maroon vinyl seat and back chrome chairs lining the wall, red fire extinguishers, a bar tucked into the corner with Captain Scott's picture above, a window, then a blank section for the escape hatch, a piano, a couple of big pictures of New Zealand's Southland, a desktop globe, reel-to-reel tape deck, a record player and vinyl LPs, a picture of the Queen beside another window, bookcases the length of the wall fronted by more maroon vinyl and chrome chairs. In the centre of the room, were the strategically placed and neatly configured art deco tables and chairs.

For the next 5½ months, this is where I'd eat, play, entertain guests and relax.

I learned during a very blunt scolding to all newcomers that morning that the bunkrooms were always to be more silent than silent and kept dark as people would come and go at all times of the day and night to sleep as their jobs allowed. The disgruntled sleeper shot me a dark scowl when I lightened the breakfast mood by venturing that the snoring I heard that morning would wake the seals in McMurdo Sound. Red-faced, I escaped to brush teeth, make my bed and tidy my room.

I stared out the window as I brushed. The earlier cloud had moved away and there it stood, the jaw-dropping magnificence of Mt Erebus (12,448 feet) with its telltale plume of steam. The world's southernmost active volcano was just as I had read, its white mantle of ice and snow cascading like some ceremonial white cloak down to the ice shelf, the centrepiece of an out-of-this world wedding.

When I looked at my desk in the 600 square foot hut I shared with the Post Office and Radio Communications, I acknowledged that life would be very, no extremely, hectic. Although I was hired by and responsible to New Zealand's Department of Scientific and Industrial Research, Antarctic Division, my costs (including salary) were shared with the New Zealand Press Association (NZPA) and the New Zealand Broadcasting Corporation (NZBC). My other roles were with the New Zealand bureaucracy – the departments of Internal and External Affairs, Tourism and Publicity and the National Film

Unit. My orientation time in Wellington also set me up to sell New Zealand-authored Antarctic books on behalf of New Zealand publishers/booksellers to visitors many of whom came from the US support and scientific community. And importantly, I was expected to take a lead role in public relations activities for visiting guests of the New Zealand Government.

After clearing a space on my corner desk, I opened my green Hermes 3000 portable typewriter. It had helped me create a lot of stories in many places in New Zealand and now it was here with me, its little lead soldiers ready to parade out into new words and phrases. The outgoing postmaster and his radio operator leaned over their counter and filled me in with what my predecessor did and what remained following his return to New Zealand six months earlier before the onset of winter. The postmaster's eyes twinkled and shone above his full black beard and, he insisted I share a beer with him. It was only 9 o'clock in the morning.

"Better than drinking juice and coffee all day. We all have one of these," he grinned, pointing to a carton of beer beside his desk. "Gotta have liquid here... very dry."

He handed over a few personal telegrams. My parents and Lois' parents wishing me a happy birthday, and a surprise message from the executive of the Taranaki Journalists' Association advising me that "Lois is in good hands." I made a note to check with the incoming Post Office guys to get on the next radio schedule to New Zealand sometime this week.

Beer drained, I went to find the other half of my job: the photographic darkroom. I turned left in the covered way and followed the duckboards uphill to the far reaches of the base. The darkroom was tucked into a corner of the 800 square foot Lab Hut. Was this where I would create fantastic historical images in black and white to withstand the test of time for the media and Antarctic Division archives? By the time I had fully rotated through the blackout door I wondered if I would ever create an image. What a mess! It had been well used by the winter-over guys. It was dark, dingy and stunk of chemicals. Ugh! It would take me at least a day to clean up, wash the trays, and locate all the developer tanks, spirals and enlarger parts. In the very dry atmosphere, the photo paper had curled and sprung outwards from their protective boxes. Much of it would have to be thrown away. Mmmm. I put spring-cleaning on the list, hoping there was fresh stuff en route.

It was small comfort to know that I was better off and better equipped than Herbert Ponting (photographer with the Scott expeditions). I could see I would have to get used to developing film and printing pictures without the aid of running water. In its place I had a chemical wash called Permafix which I could obtain from the US Navy photographers four miles away "over the hill" at the US McMurdo base. I'd never seen or used the stuff before and I have not heard of it since. It worked though for here I am some 40 years later with a moderately good collection of negative film still in moderately good condition.

Scott Base was a hive of activity with new folk coming in and the guys who had wintered over eager to get home. There were hurried, almost non-existent changeover briefings and ever-changing flight schedules due to weather. "Ya might be a bit overwhelmed...but you'll do your own thing and get the hang of it...don't sweat it...you'll do okay...good luck," were common phrases as beards headed out the door. Ten more of our team arrived in two groups over the next three days. It was tough keeping up with who was leaving. Arrivals and departures were at short notice due to the ever-changing weather conditions at the airfield. The planes landed, unloaded, refuelled, loaded and were gone again, back to New Zealand. The US base at nearby McMurdo was also changing out personnel. There were hundreds more of them to the handfuls of Kiwis coming and going to Scott Base.

In that first week, I really had little idea of how to order my day. It was head spin full on. I learned how to operate the snow vehicles I'd be using to ferry people the five miles to and from the airfield. I had to somehow organise around my communication duties by taking part in work parties and the common need to care for each other to survive, care for the base and willingly contribute as necessary to keep the base going through the busy summer and ready it for the long, isolated winter. I attempted to get control of comings and goings and to plan out the stories and pictures I would get back to the media. What would the big hits be? I wondered. I did not want to just bounce from day to day. And yet, I could feel that happening already. I raised the issue with others and while they felt much the same way, we recognised we were all in this together. I'd signed on for a rich experience in an environment that would demand more of me than I realized. The legend of Robert the Bruce, King of Scotland, came to mind. Back in the early 1300s, Bruce, suffering defeat in battle, hid in a cave and watched a spider build a web in the cave's entrance. The King's reaction had been to mimic the spider and gather fresh courage each time it fell down until the web was complete. That lesson was for me, just get up and get going.

During these first days I learned the joys of being a House Mouse – a thoroughly character forming activity to which we were all subjected in pairs and in rotation, during our tenure. In theory, we'd be on duty every 10 days. But it was not fixed. Field trips and external influences and obligations sometimes meant switching days with someone else to maintain the orderliness, cleanliness and tidiness of the base. The House Mouse was also the night watchman.

The big incentive to being House Mouse was shower and laundry time! The paucity of water meant showers were rationed until you were on duty again. And those at Base often gladly switched with "overpowering" guys coming in from weeks out in the field.

My first run as House Mouse included a work party to gather ice for the ice melters — our source of water on the base — for the lab, the galley, and the ablution huts. The melters were big heated "bathtubs" with a freezer-type door to the outside. Ice gathering was simple: hook a couple of big sledges behind a D-4 Caterpillar and head out to the tide crack where good clean ice was piled high and thick. The task then was to break the ice into manageable chunks and heave them into the sledges. Seven of us would be out for most of the afternoon. Our tools were picks, ice axes and a pneumatic spade that took a lot of starting in the minus 20°C weather and 12 mile-an-hour wind.

On my maiden outing, Doug somehow got a smack on the head with a pick. There was lots of blood and we all downed tools to have a look. Thankfully, the wound was superficial and we got back to work. In spite of the cold, we sweated up a storm inside our down jackets and windproof trousers. With the sledges filled, we headed back to base and positioned each sled outside a melter. It then became the House Mouse's job each day to fill the melters from the sledge stock.

Clearing dishes and washing up was not a problem. Collecting the garbage and dealing with it was not a problem. Stomping down the cones of contemplation (so that's why liquids were banned!) in the ablution block honey buckets (we covered with newspaper first) was not a problem. And washing the floors after supper was not a problem.

What really gave me the jimmies my first outing as a House Mouse was the night watchman bit. Everyone eventually headed for bed and left me wandering around like some friendly ghost. The responsibility called for each hut to be checked each hour – no smoke, no fire. I'd poke my head in the blacker than black sleeping huts where the rhythmic snoring was the only sound. I checked the oil levels on the Caterpillar generators. Then I headed back up the covered way and started laying tables for breakfast before dashing back around the base. Just in case I'd missed something I'd check the huts again. I was very tense this early in the game that the care of the base was in my hands. That first outing I must have gone around each hut three or four times an hour. What a way to earn a shower!

October 18, 1968

Hello My Dear:

How is it down there? Nice and brisk. Do you like being a frozen explorer? Please write me. 'Tis the only time I feel really sad, when Postie goes by and doesn't leave anything for us.

It's Saturday now and guess what? Postie brought us three lovely, oops sorry four letters. So you can take little notice of the previous page as I am happy now that I have had a letter. I know that you're very busy and your letters confirm that but I was just a bit sad as I had not even received one letter since you got there.... the girls loved their postcards. They want a board to pin them on. Next time Rachel would like huskies. They're prized possessions. You ol' sweetie for sending them. Hilary read hers all by herself.

I had to borrow some money from Mother. I had a call from the mortgage lady, so I dashed off a note and a cheque for $30. I telephoned about your pay and they said $170 had gone into the bank so now I have just finished spreading it out. Wow, no wonder you used to get brassed off doing finance!

My American ladies are coming Monday morning to begin their art lessons. After reading an article about Mike Smithers in the Daily News this morning I feel like crawling under a mat as he is an artist and me a pseudo. One of the group called yesterday for directions and just asked for the address as she had a chauffeur driven car at her disposal. Get that! I hope they don't find me too inadequate.

I had a couple of realtors up this morning for a looksee and they left with a "we'll see what we can do."

I hope your big journey goes off well. Watch the frost bite. Be careful and take good pictures so that they can colour your stories that I look forward to reading.

All my love, Sweetheart,

Tuppy

One of the books I'd send out to schools in New Zealand.

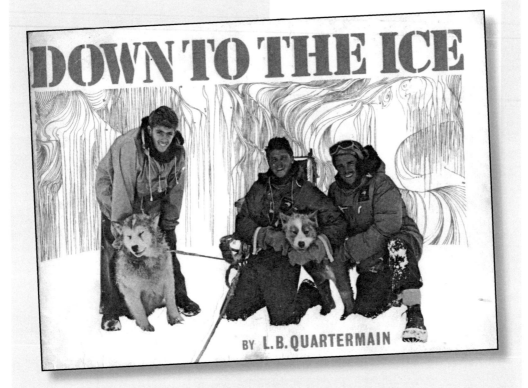

DOWN TO THE ICE

BY L.B. QUARTERMAIN

3.
A NEW LOOK AT NEWS

Room 2

Napier Intermediate School,

Jull Street, Napier

25th September, 1968

Dear Sir:

I am a form 2 pupil of Napier Intermediate School. Our class is studying the Challenge of Antarctica. We would be very grateful if you could please supply us with information your daily routine, the different occupations of the men, your recreational activities, the weather and anything else you think would interest us. Thank-you

Yours sincerely,

Lynnette Dailey.

P.S. A boy in my class would like to know if you carry (have) weapons and if so why?

(On behalf of Rm 2)

School mail was fun but I very quickly learned I would have to get a handle on it to get the backlog off my desk. It was streaming in at such a pace I figured every school in the country must have been doing an end-of-year project on Antarctica. I had not done this sort of thing before. How should

I handle it? I could not find any precedent lying around. I figured if it was me getting a reply I'd like a nice letter and a something. So that's what I did.

For the students at Napier intermediate, I sent off a couple of books which outlined the New Zealand effort as well as a typewritten letter in which I noted "...we had a blizzard last night and the wind blew at around 30 miles-an-hour. The temperature was -25°C. It is so cold at times that our breath freezes on our beards and balaclavas."

I enjoyed hand writing each response, tailored to the class age, giving a quick note on snow, ice and daylight, the scientific endeavour and the Weddell seals lounging out the front door. As with Napier, I'd pack in a couple of pictorial Antarctic books and heave the package to the Postmaster who weighed it and stamped it seconds before he cinched up the mailbag. Keeping the mail flowing with the erratic short notice plane departures was a bigger challenge than a daily newspaper deadline! At least I knew the school mail pace would slow soon with the end of the school year in mid-December.

I quickly settled into my role as the base scribe and sent out newsy dispatches to NZPA, radio and television and individual newspapers. I'd be at my typewriter most mornings tapping out 500 word pieces like the weather, team changeover, visitors, field preparations and activities, and university party arrivals and their ongoing studies. I had to take a deep breath from my former freewheeling independent news gathering training and remember that I was now working for a government agency in a remote, curious and dangerous environment. My dispatches required a sign-off by the expedition leader. For the most part this system worked very well for even this red-blooded newspaperman who had never been exposed to what might be termed "managed" news. Robin was very good and had a real sense of what could go out to the public without creating a furor in the bureaucracy. In many cases his eye on the story made it a better one and we formed a good editor-reporter team.

I had a two-tier system: short newsy pieces which could be sent by telegram and longer features and pictures which would be sent by mail to a variety of media destinations. The telegram news had to be sent by Morse code to New Zealand so I was fortunate to have the support of terrific radio operators in the Post Office.

I wasn't long into my new job when I had my first breaking news event. It was around 9.30 am and I was busy with school mail when I got word that the D-4 Caterpillar had broken through the tide crack — that place between the sea ice and the land. It was part of the landscape normally avoided and mostly out of bounds. But a couple of the guys wanted to see if the ice was stable enough to route the soon-to-be departing tractor train for Vanda on to the sea ice and avoid the long way round via the ice shelf.

I scrambled into outdoor clothes, pulled on boots, grabbed a camera, and hightailed it to the scene to capture the event on film. Driver Hugh Clarke had escaped and, terribly shaken, was OK and standing on the ice surveying the scene with others. He was a pro and had the presence of mind to drop the blade to change the weight distribution of his machine, hop out the door and on to a safe place as the ice opened up underneath his machine. The Cat dropped down to rest on the back of the cab. The ice held. It was a tense, frightening, heart-stopping moment for my new friend. The difficulty now was how to get the D-4 back on to safe ice. The rear of the tracks were in the sea. One slip and the machine would end up in a watery grave and put the summer program in jeopardy. A monster D8 Caterpillar specially lengthened and with extra wide tracks to spread its weight on the ice was working

The D4 Caterpillar breaks through the ice at the tide crack near Scott Base.

near the airfield ice road. The guys persuaded the driver of this US Navy machine to provide help. He was very reluctant but came anyway. Maybe it was the promise of a Kiwi beer. A wire rope was hitched to the D-4 and she was pulled clear to huge sighs of relief and cheering that disaster had been averted.

I was hot to trot to get this newsy little story out to New Zealand radio and newspapers. Back at my desk I wrote it, submitted it to Robin who weighed the pros and cons with his deputy, Bill Lucy. Then came the negotiations. Among the main concerns was that with the new team less than two weeks into their year such a story could be detrimental to the whole summer programme and cause unnecessary worry for those "back home". After a lot of stomping around and huffing and puffing I saw their point of view and conceded. The story was killed. It was a small but significant incident and it was too early in the season for any friction. Besides, the journalist in me rationalised, there was always the possibility of a bigger story later on.

By this time I was into my public relations role. I switched from frustrated reporter to pickup man that evening and took off across the ice highway in a Snow-Trac with trailer to pick up a couple of New Zealand Antarctic Society visitors, Frank Gurney of Christchurch and Jack Folwell of Hamilton. They arrived from New Zealand the same way I had – on Pegasus, the Super Constellation. Conditions were good and I drove out solo (in marginal and poor conditions we usually travelled in pairs) for the pickup. Tracked vehicles had their own route across the ice shelf, parallel to the ice highway used by wheeled vehicles such as the US Dodge Power Wagons, Jeep Wagoneers and our Landrover.

I loved the little red Snow-Trac. With its "little engine that could" attitude the Swedish-built, air-cooled Volkswagen-powered people mover had space for four or five. You got in and out through a tiny rear door. There was an escape hatch in the roof above the driver who had the convenience of an arm-wrestling steering wheel to tell the tracks which way to go. Passengers sat squished facing each other on either side behind the driver. You definitely had to like the person you sat next to or opposite. Scott Base had two of these little peaches that on a good day could send you over the snow and ice at a rollicking 12 miles per hour.

The Snow Trac – a Volkswagen beetle on rubber tracks.

Our guests helped load up the trailer with their gear, new supplies and mailbags from the Connie and we high-tailed it back to base shouting above the din of the engine and the clanking tracks. The two men were thrilled. It was their first time on the ice and they were here to help as volunteers with some construction work at the Vanda Station. And I had another story and picture to send out.

This was my second trip out on the ice that day, as earlier I'd been out in the Snow-Trac with base mechanic Wayne Maguiness to rescue Robin who had a flat tire on the Landrover along the ice road to the airfield. It was an eventful day spent mostly outside in -25°C with a 12 miles-an-hour wind blowing across the very exposed McMurdo Ice Shelf. I had a sensation of success and a good idea of how much could be accomplished in a polar day. It was very easy to get drawn into necessary activity around the base to exist but I still had to remain true to the work I was hired to do. I felt a little more comfortable about balancing job function with outside work parties and neato assisting-others stuff when an after meal bull session showed others faced the same challenge. It was very easy to be busy all day and not go near our primary job. Life was new and exciting and, like my new mates, I wanted to be part of it all.

"What do you do when you are not feeling well?" Lois asked me during one of our phone calls. "Maybe have another beer and keep going," I quipped.

I was in journalist heaven. There were so many feature stories and pictures bursting all around me it was a challenge to keep up. Capturing these stories of adventure, study and contribution to science meant giving up on one thing—sleep. Life was as close to 24/7 as I could get.

In a letter to my parents, I noted that in two days I might have logged five hours of bunk time and the remarkable thing was I might get a further two hours tomorrow! I was excited at being where I was, doing what I was doing. My desk was littered with scraps of paper and notebooks where I had scribbled hasty notes for expansion later into news features for the New Zealand media. I was the everywhere man, infiltrating everyone's space, taking pictures, asking questions to sift out the stories to send home. I had my co-workers' support and more often than not they would be telling me of the next "thing" they were up to. There were times I'd come back from a field trip to find notes on my typewriter "see me about...."

Leader Robin seemed to sense when the load threatened to topple me. I enjoyed it when he'd quietly drop by my desk, or maybe after coffee, and suggest "C'mon P-R-O, let's go for a walk." We'd wander outside along the tide crack and pressure ridges just talking about stuff, about home, our wives, our kids. It was a good exercise that brought things into their right perspective and priority. My sojourn on the ice was a once-in-a-lifetime event and I was determined to play my part in getting our team to the summit of Antarctic experience. Sometimes I felt quite overwhelmed by the challenges, yet I loved it. Besides I had three great reasons and their mother waiting at home to see and hear about what I was doing. My conduit to family and friends and supporters was through stories and pictures in the newspapers and on television. The opportunity I had was to keep up the flow.

The stories were there to be told. I recorded the words. After breakfast, after supper, sometimes an early morning get-up while others slept, or within an hour of the mailbag closing, I'd be at the typewriter bashing out a story. I was well used to writing to the daily newspaper deadlines and I could knock out captions and a story very quickly. My almost daily challenge was the darkroom. I groaned each time I spun the light trap door and faced into that dark space where I wanted to produce great pictures to bring a visual note to my stories.

I was used to the demands of NZPA as a result of filing reports to them and editing reports from them in my 10 years on the paper. But down on the ice, I soon learned that life would be different from anything I'd been used to.

Following my orientation in Wellington, Les Verry, NZPA's managing editor wrote:

"We shall hope to receive from you comprehensive news cover of all activities at and associated with Scott Base. We would of course like this to include from time to time reports on the work of field parties, both while they are in the field and when they return to base. The work of the various university groups is likely to be of considerable interest this season and we hope you will be able to keep us well informed as to what they are doing. We will always welcome interviews with direct quotes from expedition and group leaders when appropriate."

His letter helped me to remain focused throughout the summer as I walked the fine line serving two masters. I was both a newsman and a bureaucratic hack. In my first days, I developed a pattern of heading to the typewriter right after breakfast to clear or work on any stories I had picked up from the previous day. Late morning, I'd get nosy and wander around the base to keep abreast of what people were doing. The afternoons could be used for work parties, base activities, picture taking or soaking up knowledge about other guy's jobs. Late afternoon I might be back at the typewriter or in the darkroom.

During some of my darkroom time, I am sure the guys in the adjoining lab could hear me swearing as I tearfully struggled to make do with the equipment and curled up photo paper. Some days it was really difficult to get film into the spirals and into the developing tanks. Some of the 35mm steel spirals were bent and some of the plastic spirals were damaged. Static electricity was an enemy I confronted very early into the job when one of my exposed films slipped from my hand as I loaded a spiral. It unwound with a flash on its way to the floor. That zap was a mini bolt of lightning that ruined a whole film. Chemicals stored under the bench in the ambient temperature of the darkroom were always too warm. I compensated mostly by adjusting processing times. Otherwise I'd have to run outside, cool the liquid down (always too much) and then warm it again to the correct temperature. Adjusting processing times was a convenient solution when I was short on time. On really bad days, I only had to think of those pioneers of Antarctic photography, Herbert Ponting (on Scott's expedition) and Frank Hurley (Shackleton) and the brilliant images they produced from turn of the century equipment under the most primitive conditions. I did worry that the lack of water for rinsing my film and prints would cause them to deteriorate before they got to their destination, either in New Zealand or further afield as I responded to media and school requests.

As often as I could, I overshot pictures in the field and sent a roll of unprocessed film to some newspapers and to the Antarctic Division library. The only problem with this practice was I never saw the film again. It was great though that the pictures were used and that, after all, was the end point of my job.

For the colour slides it was a whole different story. Like anyone else, I had to send the film to a Kodak lab for processing. Then the slides would come back to Scott Base for review and I'd select the slides and send to their destinations. Again, I often sent unprocessed film to Julia Thomsen at the Antarctic Division library to short circuit the process and have her distribute the pictures on my behalf. I regretted not being able to keep a couple for my own collection.

While the darkroom was a place of frustration, I only had to think of

Lois and the day she started at the newspaper as New Zealand's first woman press photographer. Here, eight years later, the rookie was now my dark-room mentor. Lois and her former newspaper colleagues Glen Ferguson, Ray Pigney and Kirby Wright had been my fast track tutors. The Scott Base dark-room continued to make me a fast learner; when I couldn't figure something out I booked a telephone call to Lois.

I shot a great deal of my black and white pictures on the aging and battered base Rolleiflex 2¼ square camera. It was a very versatile camera and had great latitude for exposure. I kept my own Canon 35mm camera mostly for colour work, though there were times that I did shoot a 36-frame roll of black and white. The Canon was a classy little camera that produced great pictures. It was one of the very early single lens reflex cameras and boasted through-the-lens metering, a first in its day.

The Rollei was basically an upscale box camera. A twin lens reflex, you viewed and focused the picture on a ground glass screen through one lens and made the shot through the other. Then you wound the handle on the side to advance the film to the next frame. The dents and scratches on the camera showed just how robust it was. In its day, it was up there with the pros, very manageable, quality engineering with nothing to collect dust or cause light leaks. Each roll of film produced 12 pictures.

I did become good friends with the Rollei though. Apart from field work, it was very useful to take the candid black and white shots around base of our fun times when we celebrated birthdays, events or just spontaneously partied.

One story I did not get to write referred to the work of geophysicist Jim Cousins who'd been out in the field since he arrived. And the day he arrived back at Scott Base, he had time to pack and get out to the airfield for the trip home. I suggested he should call in to Rash Avery, editor of my home newspaper, and there they would extract a story and picture from him. Well, he did and Rash wrote to me to say a story about Jim's work was used on the leader page. "It seems you are enjoying yourself and by the look of recent press releases you have been doing your publicity bit for the New Zealand team."

A Rolleiflex camera similar to my Scott Base camera. This model is in my personal collection.

The newspaper ran an offbeat news snippet column a couple of times a week. All the editorial staff were expected to contribute to it. I sent one in from Scott Base and that prompted Rash to respond: " I have told the reporters that if Connell can contribute from that distance, they ought to out-pull the digit and fill the rest of the space."

The thing I missed most was that magic feeling that comes for any afternoon newspaperman when the first edition comes off the press and you see the results of your morning's effort and who made the Front Page. Now, I wrote and photographed news and feature stories that went into the great unknown and relied on Lois and others to tell me stories were appearing in the newspapers. I did not see any until I got home at the end of the summer when my Dad handed me a pile of clippings he had gleaned from a variety of newspapers that came to the office. After his work printing that day's paper was completed, Dad was a regular upstairs outside the editor's office, flipping through the exchange newspapers from around the country finding out what I was up to.

"Hey P-R-O," Foubister said to me about a week after my arrival. "You're going out on the Vanda Tractor Train!"

"I'm what?" I said. "Me....!" A true mixture of excitement, apprehension, fear and trepidation blew out of the top of my head as I absorbed the fact of being included in a key operation to the whole summer's activities for the NZARP team. We'd talked about the event often and the fact that this would be the "last of the great tractor trains". I had thought the build-up and completion of the new winter-over station at Lake Vanda would be a job for the real pros –those alpine experienced field men within our group.

I put a call in to Lois that night. "Not sure why I've been included," I said. "I'd have thought it would be the field technicians. I could be just an extra pair of hands or I could be there to write about this historical event. Over."

"How about a mixture of both," she said, ever so wisely and knowingly. "How long will you be gone?"

There was a long silence.

"You are supposed to say 'over' when you're finished a sentence, hon. The radio telephone is a one-way system."

Silence.

"You didn't say 'over'," she said. We laughed.

"Call me when you get back, over. I miss you and love you so-o-o much, over. (pause) Girls miss you, too. Over again."

I chuckled and felt a bit of water in the eyes. "Yeah," I choked. "Bye, hon, over and out." Vanda was very important to the success of our year and the whole thrust for our small party. The base and the continuation of New Zealand's scientific endeavours were in the political microscope back in the capital, and as a result, the 1968-69 team was smaller than usual. There were no government sponsored field parties this season as all resources both financial and manpower were directed at establishing Vanda Station, maintaining ongoing base studies, and providing support for the annual university programs. I'd heard the overview of the Vanda focus at the training camp and compiled a to do list of story ideas including: getting materials and supplies to Vanda, the construction and scientific purpose of the new station, its vice regal opening, and finally settling the first winter over team for the long dark night.

The decision to move large amounts of materials "by road" was simply a matter of budget plus expediency. We'd make do with what we had. The planners figured this might be a three to five day hike to get the supplies to the Wright Valley, offload them and return. I was to learn that nothing in Antarctica was normal.

The crew of the Vanda Tractor Train 1968/69. From left (standing) Bill Lucy, Hugh Clarke, Noel Wilson, Alan Magee; kneeling Allan Guard and Graeme Connell.

4.
JOURNALIST ON TRACKS

So what is a tractor train? Well, ours was comprised of:

•　　Able, a powerful, yet ungainly looking piece of South Polar travel officially known as a Sno-Cat. A 400 horsepower beast mounted on four tracked pontoons. It had been one of the main vehicles Sir Vivian Fuchs used in the crossing of Antarctica in 1957-58. It pulled a wannigan, kinda like a plywood box mini house trailer on sleds as well as two sledges loaded with equipment for the new winter-over base.

•　　Next, the NZARP D4 Caterpillar bulldozer, pulling two sledges loaded with drums of fuel and a hefty empty fuel storage tank on skids.

•　　Two Ferguson farm tractors each pulling a rubber-tyred trailer followed those two strong beasts. Specially equipped with tracks, they were later models to the veterans of the Ross Sea to South Pole leg of the Fuchs Trans Antarctic Expedition a decade earlier. The world was astounded that these vehicles, quite at home on a New Zealand dairy farm, had successfully made the journey from Scott Base to the Pole to meet up with the team crossing the continent from the Weddell Sea. Good grief, a "Fergie" was one of the first vehicles I learned to drive back in the farm days of my youth.

The Vanda Tractor Train lined up and ready to roll from Scott Base, October 1968.

Our destination was the Wright Dry Valley where it meets the Wright Lower Glacier some 70 miles northwest from Scott Base. The dry valley region includes the Taylor, and Victoria Valleys and together they represent one of the most waterless places on earth. Scott and his men first took note of the barren desert in 1903, and he sent a team including Canadian Charles "Silas" Wright there for a greater look during 1910-12 expedition. Shackleton also worked in the area during his 1909 expedition. Nothing much happened after that until the International Geophysical Year of 1957-58. That is when a couple of Victoria University (New Zealand) geology students Barrie McKelvey and Peter Webb badgered their way south and with a mixture of perseverance and good luck began probing the secrets of the Victoria Valley complex. It became a life work.

Our job included locating a safe and suitable route off the Wilson Piedmont, an extensive icefield bordering the coastline and connecting several glaciers, including our target of the Wright Lower Glacier. Once on the valley floor we'd dump our loads at the foot of the glacier, drive out and head back to base. Wheeled vehicles and tractors would haul the materials westward 20 miles up the valley to the shores of Lake Vanda, a scientific jewel in the heart of the dry valleys. The goal of our expedition year was to complete a winter-over base near the shores of the lake, adding to the facilities established in previous years to house summer field parties.

The dry valleys form the edge of the Antarctic continent at this point between the sea and the Polar Plateau. We'd be on the ice...that covered the sea ... and we'd be driving big heavy equipment on it. I heard at the Waiouru orientation that it was not known how difficult the traverse would be but the expectation then had been to head out on October 24 with completion by October 29.

Countdown for departure started in earnest on October 22. Departure had already been impacted by Hugh's escapade in the tide crack as well as locating and assembling supplies as sledges and trailers were loaded and vehicles readied for the epic trip.

I spent the morning with an ice gathering party for the melters and helping Hugh with the D4 clear snow from around the ablution block. It was a really cold morning and I froze a bit, with a bad case of frost nip to my fingers. Thawing those little lumps of dead white really, really hurt. I went into the washbasins and soaked them in lukewarm water. My fingers prickled and tingled for some time until feeling and warmth returned. Under the guise of getting organized, I cleaned out my desk after lunch to stay busy and shush my anxiety. We were ready to go, but the weather packed in and we stayed. Luckily, there was a party that night for our guest, the top brass of the Royal New Zealand Air Force, Air Vice Marshall C.A.Turner (New Plymouth Boys' High School old boy).

DAY 1

Billed as the last of the Antarctic tractor trains, we spent a good hour or so in the evening hours being farewelled for the historic journey. Cameras clicked as we gathered near the hangar and enjoyed a few cans of Leopard beer in hearty convivial Scott Base spirit. At 9 pm, Bill Lucy, our leader, fired up the Sno-Cat and with a steady rumble from its big engine led the train out, following the McMurdo ice road towards the ice runway before we swung to the north.

We were on our way. While I was thrilled to be part of this traverse into the unknown, I could not help but wonder when I would be back at base and when or how the mail piling up on my desk with each plane arrival would ever get done. The trip would be three to five days. I had no idea at all what I was in for so I tried to think in terms of storylines and pictures for the media back home and how that might be accomplished. I'd written up a story that the tractor train had begun its journey and I'd sent it out on the telegraph earlier in the day. Without knowing how the story had been featured I could only guess at how to keep the story alive. I didn't have a clue as to when my next dispatch could be made or even how. It was now Friday night. I could take pictures along the way and wrap it all up in a good feature length article when I got back Monday or Tuesday. The challenge of the journey had not sunk in.

It didn't take long to realize it was going to be cold. And I mean cold. Noel Wilson was up in Able's cab with Bill as our giant vehicle clattered across the ice. The only place for me was stretched out on a pile of supplies loaded in the back. Hugh followed in the D4 with its enclosed cab, Alan (Sam) Magee was next on a Fergie with Allan Guard bringing up the rear on his Fergie, patriotically flying his homemade "Fairlie—Gateway to the MacKenzie" flag. The tractors were open with windshields providing minimal shelter for the driver and no protection from the drift being thrown up by the tracks. Our train was strung out over about a quarter mile.

The back of Able was more like the back of a meat freezer. Lying there I was not generating any body heat and just got colder and colder, despite my cozy down clothing. I could not sit up and I couldn't see anything out the rear window. It was tough to lie there and try and see something of our journey through the front windscreen. Conversation with the guys up front was out because of the engine noise and clanking tracks. A rugby field shout worked but was punctuated by too many ehs, missed thats and wha-a-ats.

I had my thoughts to keep me company especially thoughts of the home front. I'd telephoned Lois before leaving and happily, she sounded in great spirits. I wasn't in a position tell her what day I'd be back at base, maybe

Monday, maybe Tuesday. I'd expressed some of my concerns and reservations about the whole assignment to Lois in our pre-departure phone calls.

"You'll do a good job very well as you are conscientious and versatile," Lois had written in her first letter to me at Scott Base. "It will be terrific for you and for me in a funny kind of way."

I knew that she and the girls were in good hands with her gardening, friends and family and a determination to learn to drive. She'd admitted to me she

My Canon FT camera, 1968, a state of the art single lens reflex in its time. And it is still with me and still takes great pictures (film).

was worried about overkill and really just wanted to be at home with the girls. But she knew she had to stay busy. I was certainly a very fortunate man.

Lying in the back of Able, I also worried about my Canon camera, as I'd dropped it when I was leaving the hangar. Would it still function? I wouldn't know the answer to that until I got back and processed the film. There was no option but to shoot away and trust that the camera was robust enough to withstand the knock. I was so glad I had spent our last remaining dollars in Christchurch on a small stiff-sided leather camera bag. It afforded an extra layer of protection for my gear and film supplies. I'd hummed and hahed over the purchase as the money should really have been sent back home to Lois to pay a few bills. I'd even promised her I would do that. Even though I was feeling a bit guilty, I knew the decision had been a good one now I knew the conditions I'd generally be working under.

We'd been encouraged to get personal cameras winterized to keep them from freezing. I think some of the guys had gone to this trouble but my camera advisors suggested that with proper care I'd be o.k. Lying there I developed my own cold weather strategy to avoid any danger from condensation when coming in from the freezing cold to the warmth of the buildings. The bag was a big help with this and I allowed the camera to gradually come to room temperature. This often meant leaving the camera in the bag on the floor of the sledge room or similar unheated or low heated area of the base to acclimatise before taking it to my office or darkroom. In the field I kept it away from all sources of heat.

The Sno-Cat rumbled along at about 10 miles an hour, with the wannigan and two supply sledges sliding along in tow. About an hour into the journey we stopped to wait for the D4 and the tractors. Allan had spark plug trouble with his Fergie and he (embarrassingly, the base engineer) had to borrow tools from some nearby Americans. His toolbox was aboard the Sno-Cat and as we'd not really talked about how the journey would proceed, Allan had expected all the vehicles to travel close together for safety and not get spaced miles apart.

This stop allowed time to make up a brew of soup in the wannigan before we pushed on to a US science research seal hut that provided convenient hostelry for the night. At 1.30 am, it made for easy camping. We'd covered 20 miles since leaving base. Mmmm, only 60 or so miles to go! The seal hut was nothing more than a yellow painted plywood shack with a great hole in the floor over an equally great hole in the ice to provide access for divers to the sea and whatever science guys wanted to get into or out of the murky depths below. The hole was covered with a loose plywood sheet. Mmmmm, fascinating. I'd be sleeping on that! Ice and ocean just inches away!

"How long does it take to freeze if you fall in? " I asked lightheartedly of no-one in particular.

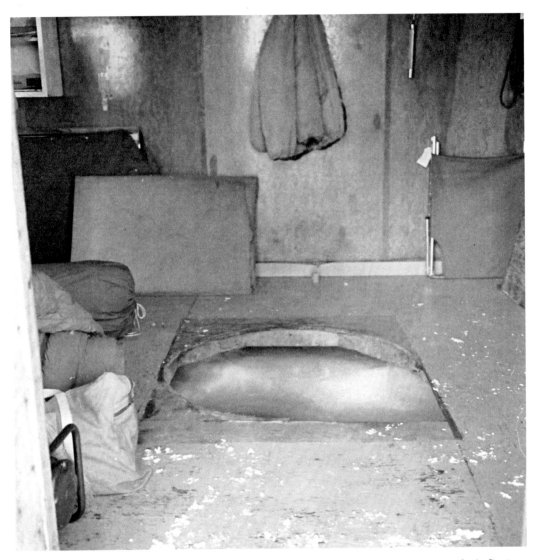

The hole in the floor of the seal hut where the Tractor Train boys spent their first night of their traverse. A piece of plywood covered the hole, I rolled out my sleeping bag on it and snored the night away.

"Oh, it's not that P-R-O; you'd last a few minutes," Bill replied. "The trick is finding the hole again when you come up for air because the current carries you away! I reckon you'd drown before you froze." There were big chuckles all round at Bill's very matter-of-fact response.

With a seasoned explorer's confidence, I unpacked and rolled out my bedroll for the first time, quietly watching the others to learn the bedtime skills of being in the field. Then I realized four of us were in this for the first time. My toes were frozen. I took my socks off and stared at the numb white lumps attached to the ends of my feet. Wiggling them and rubbing them, I painfully thawed them out. Owww! They hurt as life spread into them. It was something I was a bit used to as I always ended up with numb toes and hands after a long training workout at the swimming pool. Before long I was snug in my heavy down sleeping bag, comfortable and warm and dreamily joined in an unholy, unharmonious first night snoring chorus.

DAY 2

"Hey, you guys goin' t'be here long?" It was an unscheduled 10am wake-up call by a couple of jovial American scientists. We were in their hut. But they were not in a hurry for their hut as they had to go away somewhere in the pursuit of science and would be back. We had plenty of time. We shook ourselves wide-awake and started a morning routine of getting out of snug warm sleeping bags. The challenge was simply to avoid the sleeping bag's ice encrusted opening where my breath had frozen. Sleeping in polar mummy bags, you would get in almost fully clothed and slowly peel off as the frozen bag warmed up from body heat. By wakeup/get-up time I was almost naked and as warm as just-popped toast. Once I'd gathered my breath to emerge past the icy opening, I had to fidget in the foot of the bag to retrieve my nice warm clothing. In the end there was no easy way; just take a deep breath and jump to.

This was our first get-up of the trip. It took a while even though we were eager to get on with the adventure. We packed up from our hut quarters, refuelled the vehicles and breakfasted on coffee, scrambled eggs, bacon and stewed apples, a royal breakfast under the circumstances. By this time our hosts had returned with a beautiful silvery, spotted Weddell seal they had picked up at Cape Armitage. He'd be measured, weighed, tagged and, with a tracking instrument attached to his back, released back to the Ross Sea of McMurdo Sound through the hole in the ice under the floor of the hut. Science at work in the polar latitudes.

It was after midday by the time we had finally packed up and had the vehicles rumbling in readiness. Our American hosts gave us a great send off – a motley crew of Kiwis and their low-cost, ultra-used "she'll be right, mate" supply train. Our goal now was to negotiate the expanse of ice ahead. It was vast. It seemed all sky. Direction? Keep Erebus behind us to the right, more or less southeast.

We drove on and on, clanking our way over the ice and snow at maybe four or five miles an hour. The further we traveled, the more ruffled the surface as we headed into the full open expanse of McMurdo Sound. The surface had definitely not appeared lumpy and bumpy in the pictures I'd seen. But here at ground ice, it was humped and hillocked by sastrugi, the name for hard, crusty windblown rifts maybe up to two or three feet high. Travelling on this surface fascinated me and each time we stopped I found a new patch to try for the definitive photograph. I figured the only way to put this on film lay in how I used the light. So I shot frame after frame of black and white film, sidelight, backlight, front light, any-which-way light. I used different camera settings (aperture and speed) and played with the orange filter, hoping to see success when I reached the darkroom.

Travel in this part of the world is basically by dead reckoning. Very little in the way of landscape features to work from and a compass is of little to no use because of the proximity and influence of the south magnetic pole. Bill

The sastrugi covered sea ice of McMurdo Sound.

was our pilot and we all just figured he knew where we were going. After all, he'd been to this part of the world before. Being a surveyor by trade might count, I thought. I was part of a quartet (Hugh, Noel, Allan and myself) of youngbloods plucked from the calm, peaceful suburbia of small-town New Zealand playing follow the leaders (Bill and Sam).

Later in the afternoon I relinquished spectator duties and took the wheel of one of the Fergies. Noel and I drove the tractors for about 21 miles. I'm not sure about Noel but I just went into a daydreaming trance. Pictures of Sir Edmund Hillary and his team driving Fergies on the polar plateau were implanted in my brain. Now here I was clattering across the sastrugi landscape in charge of a farm tractor. It took me back as the Fergie (albeit an earlier model) was almost the first vehicle I learned to drive in my very early farm days. As an aside, the first vehicle I drove was a Model T Ford. And I was only about eight. Drive is probably a stretch but I drove it through the farm gate while the owner shut the gate. He then allowed me to pilot it up to the house. Next stop the Fergie.

Bill Lucy checks the ice of McMurdo Sound to ensure safe travel.

Allan and Sam joined Hugh or Bill to warm up in the sheltered, enclosed cabs of the D4 and Sno-Cat. The tractors just had a basic canvas windshield. When not driving, I found it was more comfortable and slightly less chilling to ride standing on the tractor drawbar with heavily gloved hands clutching the rear wheel guards as we rumbled along in seven-mile hops. While the driver was exposed to the wind and drifting snow whipped up by the tracks, I was sheltered in behind. The Sno-Cat would wait for the others to catch up. The D4 was the slowest.

Every now and then, Bill would halt the train, clamber out and wander around looking over the sastrugi for any sign of tracks that might still be visible in the wind-driven snow. These tracks would be the remnant of a spring reconnaissance trip some six weeks earlier by the outgoing winter-over party as preparatory work for our push with materials to upgrade Vanda Station. The scene was unchanging as our train clanked on across the white, always white, expanse. I deferred a news report for the New Zealand media as we expected to be at the next staging point in the Bay of Sails within 24 hours. I wrote a brief media report in my head, outlining the first day's progress but left it there as there was little of substance to report, just mile after mile of rattling vehicle tracks and strained eyes from staring into the white, scanning the way ahead for any sign of distress in the surface. We kept this up for six hours, until our stomachs decided it was time for a halt and food.

The beef stew cooked up on our campstove was just the ticket. Nourishing for body and soul. We ate in the lee of the vehicles, marveling at our surroundings and gazing out over the white landscape knowing that we had many miles yet to traverse. Though we were well into the evening, we agreed with Bill to plug on and try and make the Wilson Piedmont before bedding down for a rest. With virtually 24 hours of daylight at these deep southern latitudes, it did not seem out of the ordinary. We'd all grown up under the mantra of work with the light and sleep with the dark.

We pushed on, just thrilled with the fair weather. However, the going became more difficult with higher, lumpier sastrugi and greater exposure to the wind the further we traveled away from the influence of Ross Island. We were making more stops now with the Sno-Cat coming to a halt and waiting for the D4 to catch up and knock a path through the sastrugi for easier passage. Able was feeling his age a bit, suffering from badly worn links on the rear pontoon tracks. One of the tractors struck a spot of bother when a pin on the drawbar sheared off and we had to stop and replace.

Nervous excitement kept me wide-eyed as the splendour and grandeur of my surroundings expanded with every clank of our vehicle tracks. There, on an ice platform in the middle of the sea miles from the shore, colours unfolded around me. Light greens, pinks, white ice, green ice, blue ice, big chunks

Sastrugi on the fast ice of the Bay of Sails as the train approaches
the Wilson Piedmont Glacier.

and little chunks. And would I ever forget the grandeur of a mid-afternoon parhelion, that mixture of sun and ice crystals forming mini rainbows above, below and either side of the sun. I thought of ways to describe this to Lois who loves colour and who has the wonderful ability to paint all the subtleties of tone in a flower. It was like a heavenly compass pointing the way forward.

By midnight we reckoned we'd covered a good 50 miles. It was a blessing, too, that around this time we picked up scattered signs of the reconnaissance party tracks, a relief and big help as we rounded into the Bay of Sails. High sastrugi ridges continued to slow us down, and again we found ourselves travelling at less than a mile an hour. We watched the sun set and rise again without any real change in light. Our train was moving in a wide arc now, grinding round from the sea ice into the large bay and towards the shoreline of the next challenge, negotiating the way up the Wilson Piedmont glacier from the sea ice in our push towards the Wright Lower Glacier and entry to the valley. I had no idea of the temperature, maybe -20°C to -30°C with constant wind making it even colder. Our reddish-orange down clothing, multiple layers, movement and food kept us warm. We watched each other's faces for any trace of white, a precursor to frostbite, our hands firmly garaged in giant, furry bearpaw mitts, our feet protected in heavy woollen socks inside bright yellow polar mukluks.

The Bay of Sails, remarkable in the midnight twilight, is named for the large number of huge, towering icebergs stranded and locked in the sea ice – trapped in time, patiently waiting for the next thaw to continue their odyssey to the open sea. Around 4am, bathed in that strange "daylight" washing of pinks and

An iceberg in the Bay of Sails.

blues over the "sails", we finally reached the pressure ridges of the tide crack marking the shoreline between the frozen land mass and frozen sea, where tremendous unseen forces had surged and buckled the ice.

We'd been on the go for more than 16 hours, fighting the terrain, mechanical challenges, the incessant grind of the vehicles, bitter cold, and the constant peering into the white nothingness. My ultra fatigue was overcome by the brilliant play of light in my surroundings and the approaching glacier. I drove a Fergie the final six miles to the tide crack, thrilled and anxious, paradoxically happy and excited at being close to the point of danger. Everywhere I looked, I could see cracks in the ice: leads, I was told. We'd stop and check and look for best place to cross at right angles. At times we followed alongside a lead looking for the crossing point. Would my tractor tumble through a suddenly widening crack into a watery end? The sight of Hugh's brush with disaster in the sea ice at Scott Base was still crystal. Gale force winds the past few weeks had removed any trace of earlier tracks. We were on our own once more.

Some 50 yards from where the glacier met the sea ice, our crazy train came to a halt. This would be a good spot for a meal and while that was being prepared, I went with the experts to find the safest and easiest place to move through the jumbled ice at the tide crack. Ice axes probed for soft spots as we fanned out testing the footing, not only for our train but also for the supply runs that would follow.

A cold and lonely traverse in the late evening October twilight.

We decided to stop. This would be Camp 2, right here on the sea ice. "Bloody hell," I muttered quietly to myself. "We're almost on the edge." My jitters must have shown as a voice from the other side of the sledge yelled out: "She'll be right P-R-O. Safe as a rock." Noel was on the job though, wandering around with a probe to ensure safe thickness of ice.

The polar tent was unshackled from the sledge and the purplish heavy canvas pyramid was soon erected a few yards from the wannigan door. It was guyed into place and ice and snow piled around the edges. I had drawn a berth in the tent with Allan and Noel. Bill and Sam would bunk in the wannigan and Hugh rolled out his bag in the back end of the Sno-Cat. I loved tenting. I'd carried my liking and Boy Scout training through to our married life. But tenting in these temperatures in an historic looking tent here in Antarctica was certainly a 10 on the camping scale.

Our tent looked like the ones in the Scott movie of my youth. I looked at it and thought of the polar pioneers and quietly wondered what sort of a story this tent had to tell. I retrieved my bedroll from the back of the Sno-Cat and tried to shove it through the hole in the front of the tent, only to get it tangled in the "tunnel" or sleeve of yard-long canvas chute that got tied off on the inside to keep the Antarctic winds and snow out. Mmmm, I thought, how do three men get outta there in a hurry?

I discovered the knack of entry soon enough, for one person anyway. Crawl in head first on my hands and knees, stop halfway, bang my feet together outside to free loose snow and ice from my boots and drag my legs through the tunnel and kinda roll over at the same time to be sitting up inside. To get out much the same: kneel down and wriggle my arms and head first into the outside world. I learned to put mitts on pre-wriggle.

The Stew a la Sails (we really just heaped a whole lot of frozen stuff into the pot) could never be reinvented. It was definitely a guys' meal. Its beauty lay in the hole it filled and the warmth it provided. I dined well and washed it all down with coffee and a beer and headed for the sleeping bag. It was something like 6 am and I was emotionally and physically spent. Who cared if I was sleeping on a piece of canvas laid out on ice just inches above the ocean? I rolled up my down parka for a pillow, took off my boots (they'd be frozen in the morning) and put them to the side of the tent and tucked the felt liners into the sleeping bag. Next I piled in, fully dressed, but shivering. I wiggled down and pulled the hood over my head, balaclava included. By the time Noel and Allan wriggled through the hole, I was warming up and starting to strip to my underwear. I started drifting towards sleep thinking about Captain Scott and his men and how they ever kept warm in wet skin sleeping bags. At least I was dry and my bag was top quality down. I disappeared into the snore zone as my chattering companions went through the rituals of getting to bed.

I came to some hours later because I was hot and had to shed more clothes. It did not pay to move too much, though, as shards of frozen breath from around my face hole found their way into the snugness of my bag. Our combined breathing could be seen in the eerie light clinging to the sides of the tent. I was now stripped to my underpants, warm as toast and, with the ice brushed off my bag, I curled up and closed my eyes. Sleep, after a rest, was harder to come by now and I just lay there listening to the grinding of the ice beneath the thin canvas tent floor. The wind howled its siren song outside and the diesel-powered D4 Caterpillar underwrote the score with its base tones. We had to leave the Cat idling at all times to keep the engine warm. If it was turned off for any period of time we would not be able to restart it. The petrol (gas) powered tractors and Sno-Cat were different and, while difficult, were easier to start in the vicious cold. And our two mechanics, Sam and Allan had their ways...

DAY 3

Wakeup time came and I just lay there wide-eyed, marvelling at my first night of tenting on sea ice in the howl of the Antarctic wind. I looked around

Our polar tent home pitched near the tide crack at the foot of the Wilson Piedmont Glacier.

the tent from my wee frosted peephole and wondered about the time of day or night and whether my companions were awake. With rude grumblings they told me they were. We spent a few minutes talking about the timing of who did what first, and the best way to get out of the bag and get dressed in the restricted space. At issue was how to avoid the ice of frozen breath around the mouth of the bag and how to prevent touching the tent walls and raining icy crystals down on everyone. I'm not sure if there was any sort of a vote, but I was the first to leave the bag. Nature's call was very loud and clear. There was only one way, get out of the bag quickly and move fast. I started fishing clothing from inside the bag and carefully unzipped and folded the bag back to avoid as much ice as I could. Then shivering, and to the snorts of tent mates, I untangled my clothing, turned it right side out and dressed. Getting dressed was one thing. Getting my boots on was another before clambering over a body, untying the door tunnel and crawling out. Daylight white hit hard and the bitter wind provided the double whammy before I could wrap myself in my down parka.

The important thing now was to run around into the lee of the big sledge so I could relieve myself as quickly as possible. Let's see now: woollen work trousers, woollen long johns, string cell underwear, briefs. Ahhhh! Somehow, this was more calculated from any of our earlier nature breaks. We'd decided earlier in the trip to count off our clothing after Bill recounted the story of a poor fellow in a hurry who forgot one layer and ended up peeing his pants.

By the time all six of us had gone through this start-of-day ritual, we realized it was after 6pm. We had slept through 12 hours. It was night again. After packing up the tent and our gear, we gathered in the wannigan for food of the nothing-remarkable-but-oatmeal-is-good variety and continued to plan for crossing the dangerous tide crack. Even though we had full light of day, we decided it was far too cold to start out and it would be warmer and far more sensible to travel in daytime hours.

We re-erected the tent in a different spot and "remade" our beds. I was relieved we did as Noel, probing around with his ice axe, found a four-inch wide crack right under where the tent had been the sleep before. Was that the loud cra-a-ck I heard in the night?

The temperatures had dropped from yesterday's minus 20°C to around the minus 30°C mark but the wind chill made it much colder and more difficult if we ran into trouble. As we sat around yarning over our meal of the usual trail stew, I felt a lot more comfortable as I grew into the strange, silent and fascinatingly picturesque environment. It was a good time for a cigarette or two, chocolate, a Leopard lager or three and coffee.

Somewhere in our joking around Hugh reckoned I resembled Captain Oates of the Scott party because of the way I wore my green balaclava rolled up at the rim and low down above my eyes. Every now and then I'd hear, "Hey Oates" and knew it would be me. The nickname that had stuck from the day we arrived on the ice was P-R-O. I was known more by that than my own name, in the same way the carpenter became Chippy, the chef Cookie, and the Welshman, Taffy.

Given the weather-induced change of pace, I spent about half an hour scribbling out a news report. The item detailed the traverse so far and our safe trip across the sea ice of McMurdo Sound. The brief 300-word report gave an account of how we were now ready for the push up and across the piedmont glacier to the Wright Valley. Sending out a news dispatch from the Bay of Sails, or any field location, was truly an exercise in teamwork. We were all huddled in the wannigan after our meal. We spent some of the wait for the scheduled midnight radio contact with Scott Base wordsmithing the news item. Bill was our Morse code operator, and he used a field Morse key strapped to his knee for steadiness. Bill filed his field report and then letter-by-letter tapped my story through to the radio operator. Fifteen minutes later he signed off. My first-ever Morse code news report was on its way. The base operator would now transcribe and forward it to Robin for his approval the next morning. Once he got the official go-ahead, the radio operator would then send the dispatch via Morse code to an operator at the New Zealand Post Office in Wellington who would transcribe it and deliver it to the New Zealand Press Association and New Zealand Broadcasting Corporation. These agencies would then distribute the report via telex machine to newspapers and radio stations throughout the country.

All field parties had radiotelephone links back to Scott Base, but more often than not atmospheric conditions in our year prevented voice calls. Solar flares (or sun spots), and I say this very unscientifically, ruffle the "nice even" curve of the ionosphere way up there outside the curve of the earth's surface. Radio signals like a "nice even" curve to zigzag (reflect) their way to where they are going. The sunspots mess things up and cause the radio signals to pout: scatter and cause disarray to the angles. No "nice even" reflection off the ionosphere meant no reliable radio communication whether on land or in the air.

DAY 4

We awoke to a rare day Saturday, October 26. I lay in the eerie light of our double-walled canvas tent wondering what was different. No clues. My tent mates sensed the same thing, but all we could hear was the low rumble of the D4. I offered that perhaps we had broken away and were magically floating out to sea on an ice floe. I scrambled about knocking ice off the tent walls, dressed and poked my head out the tunnel.

"No wind," I bellowed back.

"Then get your boots off my bed!" Noel rumbled in his early morning gotta-get-outta-bed voice.

It felt glorious outside, sunny and clear, and just great for our assault on the glacier. Our instinct to hunker down during the cold and return to daytime hours was right. It was 10:30 am by the time we packed away our camp. Thoroughly refreshed and with a good feed we fired up the vehicles and began to roll ahead. The Sno-Cat moved a few yards before there was an almighty graunch from the back of the 12-year-old workhorse.

"Oh, no-o," "Oh, shit!", "Bloody hell", were just a few of the assorted Kiwi expletives that flew around as we looked at broken track links and pins. Gloom settled over the train and Sam and Allan muttered words their mothers wouldn't recognize as they set to work with frozen tools on the frozen landscape to do what they could to get Able rolling again. Bill was doubly mad because he could not raise a response on the Scott Base emergency radio schedule to report our predicament.

I have no idea what they did, but our engineers made repairs enough to get us going again. Spirits rose. We needed that to get across the tide crack and onto the glacier. I confess to being extremely tense as our vehicles eased one by one from the sea ice to the shoreline. I hung back and shot off some film and then jumped on a tractor drawbar. Everything went well and we had a good crossing. The D4 incident back at Scott Base showed just how vulnerable we were and how careful and watchful we had to be. I surprised myself at the depth of my own anxiety. I'd always figured I was a bit of a risk taker. I'd been around boats, mountains, the open sea, rivers and lakes but this approach to the unknown was something new. It was silent and chest tightening. It was up there with the nerves you get maybe with public speaking, or getting to the church to get married or the birth of a child. So I changed thoughts and put my mind to thinking about Lois and the girls coping on

Approaching the climb up the coastal face of the Wilson Piedmont Glacier.

our slender finances. Even though I knew she was in good hands and very capable, I had to get over my home concerns and get used to the perils of my new environment, as this would be my life for many weeks to come.

We rumbled forward, thankful for our mechanics. Vehicle by vehicle, the train eased across the tide crack and regrouped at the foot of the glacier. The way ahead appeared straightforward even though it was going to be a stiff climb up the glacier rising about 600 feet to the plateau. I rode on the drawbar with Allan so I could get on and off to take pictures, warmer and more comfortable than on the Sno-Cat that was climbing steadily ahead of us. Hugh was not far behind with the the D4.

About half way up, the Caterpillar started digging down and skidding in the light snow cover. I left the tractor to go help. We uncoupled the sledges so the D4 could climb ahead unencumbered to a flatish point. The Sno-Cat slipped its tow and returned and hauled the D4 sledges one at a time. This leapfrog style worked. Meanwhile the two tractors had forged ahead, coping with the snowdrifts and ice. Sam wanted to avoid the icy slope and piloted his Ferguson over to a strip of rocky moraine to the south and was carefully picking his way up.

Just when we thought we'd make it to the top in one piece, a heavily-laden sledge broke free when the drawbar on the Sno-Cat snapped. Oh, man! All we could do was watch helplessly as the sledge gathered momentum and whizzed down the slope to the tide crack. When the runaway stopped, I'm sure our sighs of relief and words of utter frustration could be heard back at Scott Base.

The drawbar had fractured right at the neck of the shaft that allowed the twist between the vehicle and the sledge. Group talk arrived at the creative solution of wrapping a six-foot wire rope strop around the towing frame and securing it to the sledge with a D shackle. Man, I thought, these guys are good. Their calm and practical approach gave me a new sense of confidence. The wayward sledge was retrieved and brought limping to the top of the glacier.

We were not out of the woods yet. With morale boosting banter, we unpacked the sledge and redistributed part of its load amongst the other sledges. Then about three feet of damaged runner were cut away at the rear and the sledge reloaded. By now we were all pretty exhausted. With evening settling in, we had a sweet view across the 12 miles of the Wilson Piedmont towards our destination at the base of Mt Newall (6300 ft) at the southern edge of the Wright Valley. Over to the southwest, King Pin (2500 ft) stuck his nose, isolated and alone, above the icescape. It had been a gruelling day. This exposed, windswept crest of the piedmont became our Camp 3. A sombre crew gathered in the wannigan that night. The expectation back in New Zealand was that the whole traverse had been planned to take three to five days. We were already at that marker and while the destination was in sight I knew there were heaps of the unknown ahead. Today's mileage was basically a fat zero. Food, warmth and a chat lifted our spirits enough to go to bed.

A US Navy Sikorsky helicopter drops spare parts and supplies to our stricken party.

DAY 5

A bright orange US Navy Sikorsky helicopter announced morning as it approached for a landing to drop off a replacement suspension spring for the troubled Sno-Cat. Our contact with civilization was brief. The chopper dropped in en route to another mission. We exchanged pleasantries; the loadmaster was back on board, the door closed and up and away she went leaving us in the silence of our unlisted campground. Another example of the Antarctic frontier, I thought. Weather and radio conditions could change at any moment and the helicopters, as a general rule, did not fly without radio contact to McMurdo. We were alone again, left to our own devices and the mechanical expertise of Sam and Allan, the creative practicality of Hugh, the polar experience of Bill, and the alpine and do-it-yourself approach of carpenter Noel. Me? I could operate a typewriter and read upside down in reverse (newspaper style). Not a heck of a lot of use out here except as a tractor driver. My contribution became my hands, my lateral thinking and perhaps a minor role as cheerleader.

Our diversionary chatter included refreshing our minds back some 50 years or so when six of Scott's men known as the Northern Party spent almost a year in the area and along this coastline with nothing but their ingenuity and remarkable sense of survival. No radio. No helicopter to hustle in some piece of equipment. No chance of rescue as winter closed in on them. They dug a cave in the ice on Inexpressible Island way to the north of where we were and waited for spring, reliant on the meagre supplies they had and any available penguins and seals. Their mode of transport was themselves, hauling all their gear on heavily-laden sledges.

Once more, our fix-it geniuses burrowed under the Sno-Cat to make repairs. It was fiercely cold and they laboured in slow motion, flat on their backs in the snow, mittened fingers doing what they could, bare fingers needed much of the time. I was one of the idle, helpless ones who engaged others to keep warm by chopping snow blocks from the glacier to build an igloo. We weren't very successful. But at least we were occupied and warm.

The biting southerly blew straight from the Pole itself with seemingly nothing in its way to alter its chilling journey towards New Zealand. I was up to six pairs of pants: briefs, cell underwear, woollen long johns, woollen trousers, down pants and orange canvas outer pants to cut the wind which finds its way through anything. On the top half, I was protected by my cell undershirt, woollen long johns, woollen plaid shirt, green wool sweater and a heavy orange ¾ length down parka with a fur trimmed hood pulled over my rolled up green balaclava. To hold out the chill around my neck, I wound a very patriotic black and gold scarf of my home province of Taranaki. My younger sister Fiona had knitted it especially for the great polar adventure.

We kept our eyes on each other's faces and noses for any sign of white skin indicating frostnip. When we did see those telltale patches, gentle massage with a warm finger restored life to the freezing skin.

After warming our frozen engineers with coffee and trail biscuits we rumbled off across the gradual uphill climb. For around six hours, we trundled on trouble-free through soft snowdrifts and over glare ice and snow drifts towards the confluence the Newall Glacier, the Wright Lower Glacier and the piedmont at the entrance to the Lower Wright Valley. With open leads on the sea ice behind us, our concern now was for crevasses. I kept a wary eye on the snowscape from my drawbar position for any slight depression in the surface that might indicate a crevasse. We did not want any more incidents.

It was just after 7 pm when we arrived at our destination. This was Camp

4, roughly 1300 feet above sea level. The cloud had wrapped around us and it was bitterly cold. We set up camp and went for a body-warming walk for a good look at the terrain where, we supposed and had been told, a track could be formed for the tractor train to get off the glacier and down about 300 feet or so to the valley floor.

Barring the drop into the valley, the tractor train had all but reached its destination. The big plan in the establishment of the Vanda Station called for our load of supplies and materials to be dumped at the foot of the glacier

The tractor train heads west across the wide open Wilson Piedmont Glacier to the Wright Valley.

so that later parties could use wheeled vehicles to transfer the supplies some 20 miles up the valley to the Lake Vanda site. Our next major task was to get to the foot of the glacier. The intention was for Hugh in his D4 Caterpillar to blade a track off the Wright Lower Glacier. The "where" was the big unknown.

After setting up camp, we hiked down to the valley floor. To the west the valley opened up in huge tones of brown and grey and I understood why it was always likened to Mars. Just rock, sandstone, gritty, cold, huge, wind-swept, ginormous... and yet totally magnificent. It swept away in front of us as we wandered around. This valley stretches for around 50 miles west to the Wright Upper Glacier, which then lifts to the polar plateau at 6000 feet. In parts, it is more than five miles wide and the barren mountains rise up to around 6000 feet on either side. To the east, in the subdued evening light, blues, greens and pinks glittered on the towering glacier wall rising a neck bending, open-mouthed 250-300 feet above the rock and scoria mosaic of the valley floor. I knew I'd have to get back next day for photographs when better light might show off the colours.

What appeared as a ready-made track led down from a neve to the south and off the slopes of Mt Newall's eastern face. It looked like we were on the right approach. The trick now was to get the vehicles and sledges off the ice and over a rocky ice hump near the top. With this approach decided we relaxed and wandered around the valley at the base of the glacier picking over rocks and basking in the might of this strange and wondrous landscape. In the centre of the valley at this point is the Onyx River, the largest watercourse in Antarctica. I only saw the dry riverbed as the river only flows in a couple of summer months carrying meltwater from Lake Brownworth and the Wright Lower Glacier on the northern side. The valley floor is some 1000 feet above sea level here and the river flows west away from the coast to Lake Vanda that is around 400 feet above sea level.

In our meanderings, we came across a mummified seal. Bill told us there were a number of studies into how and why the seals got up into the valley. One study, he said, indicated the seals were remnants of when the sea flooded this far inland. And goodness knows how long ago that was. Rule of the region: leave in place. No wonder this fellow was mummified as this is the coldest, driest, windiest place on earth.

Back at camp we celebrated our arrival with another one of our famous meals from a pile of cans, developing yet another meaning to the term pot-luck. This time we added plastic spuds, the variety of potato in a bag reconstituted to a mush with boiling water and laced with pure New Zealand butter. We agreed it was a great meal "to put a lining on your stomach, P-R-O." We washed that lot down with a Leopard and finished with a cigarette. Life seemed good again. We even speculated that we'd be off the glacier tomorrow, dump our loads on the valley floor and be heading home.

The neve rose up a steep 400 feet rise in elevation above our camp. It looked like it would give good access to the rocky icy hump we had to get over on to the rough track. Bill was gazing up to steep slope and a couple of minutes later declared, "Reckon I'd like to take old Able up there without the sledges and check what the going is like. What about it, Hugh? Follow me up in the D4."

Off they went while Noel, Sam, Allan and I stayed warm inside the wannigan fixing up a pot of tea. We watched their progress out the window as they slowly faded to near ant size. From where we were it did not seem a big task. Their rate of travel indicated that the surface must be reasonable. The Sno-Cat, as the faster of the pair, was doing well and arced over towards the valley rim where we all considered a "natural road grade" would lead across the scoria and down to the valley floor.

"Yeah, looks like they're gonna make it o.k.," Allan said. "If the track looks good we'll be able to get the sledges up there, unload, and then we're done."

With the big Sno-Cat at its destination we focussed on the D4 as it began the arc in the Sno-Cat tracks. I was busy pumping the campstove for our tea when I heard the yelling. "She-it! Hugh's gone in."

The author inspects a mummified seal found in the Wright Valley near the foot of the Wright Lower Glacier.

We crowded the window. I burned my hand on the stove. The eight-ton Cat had lurched backwards into a crevasse. Allan snatched the binoculars and seconds ticked like hours as we peered out the window waiting. The D4 did not move and we breathed relief when a barely visible Hugh came from around the front of the machine. He was safe.

With this drama, Bill turned the Sno-Cat around and retraced his tracks towards the D4. Fifty yards from the bulldozer the Sno-Cat tilted and stopped. We could not see what had happened or why Bill stopped and clambered out of the cab. "He must have gone in," Allan said, following the action with the binoculars. "Looks like one of the pontoons hit a slot..."

No-one spoke. I shut down the stove. Almost silently and without much comment we pulled on our polar clothing. I shouldered my camera bag; we gathered ice axes, shovels and ropes and began the stiff 1½ mile climb. About halfway, we met Bill and Hugh who was carrying his bedroll and gear as his Sno-Cat bedroom was no longer available. Bill had his camera and I think he was more rattled by the possibility of that being damaged with the jolt and crash to the floor from the fall than he was with the pickle we now found ourselves in.

"Bugger it," said Bill as he met up with us. "The D4 is wedged in a deep hole with the tracks on one side of a flamin' crevasse and the back of the cab on the other. Sno-Cat slotted a pontoon. Shit!"

Hugh added that as soon as he felt the rear end tilt he dropped the blade to change the weight. "I think that stopped her from going right over," he said. "Big hole and I had to be careful getting out."

We consoled each other the rest of the way down the hill, very thankful that Hugh, though shaken, was

Our D4 Caterpillar lodged in a crevasse high up a neve on the slopes of Mt Newall, above the Wright Lower Glacier.

ok. He was an experienced heavy equipment operator and had reacted with true skill. With two strikes into crevasses in as many weeks he would be known good naturedly, as "slot" Clarke. What a man!

Once again, it was a pretty sombre group that hit their beds around 3 am. Hugh joined us in the tent, the four neophytes together, fresh from the streets of downtown New Zealand. Earlier, when we first set up camp, we had poked and prodded around the tent to ensure were not tenting on a crevasse. I drifted to sleep to the grinding and crackling of ice beneath the tent's thin canvas floor. The sound echoing underneath me was different than our other camps and there was a sense of movement. I was glad to have the company, good humour and friendship of my tent mates. I can now personally attest to the fact that glaciers move.

Noel Wilson uses an ice axe to probe around the glacier just in case there's a crevasse lurking under a snow bridge.

5.

A TRIP EXTENDED

DAY 6

We woke around 11am and, after the usual stomach filling feed, headed up the hill to inspect the vehicles and assess what we had to deal with in this latest round of mishaps. Under Noel's guidance, we poked and prodded our way around the stricken vehicles finding the extent of the crevasse field. Each crevasse was flagged to make sure none of us broke through the snow bridges and tumbled in. I took pictures of everything and roped up and dropped down the crevasse under the Sno-Cat. I was down only a few feet, but man, was it cold. I was after both the necessary archive pictures as well as a chance at some pretties for myself. The crevasse dropped way down below me. There was no bottom, just dark blue fading to black. It was huge frozen cavern that anyone falling through would likely not return. I think we all had a turn on the rope and were thankful it was the Sno-Cat in the slot and not one of us. Wedged into the maw of the crevasse, the D4 was not going anywhere. The vehicles lay silent and cold. Starting the diesel D4 would be a challenge once we got it free.

We plodded back down the hill to our little camp and gathered in the wannigan. I listened as the others chattered about the predicament of the vehicles and how to extract them. They drew diagrams and bantered back and forth on how six men would use their hands and ingenuity in place of cranes and tow trucks. Their thinking process and sharing of ideas was a pleasure to listen to. From this wonderful meeting of practical and creative brainstorming they made out a list of needs for Scott Base to send in. I scribbled out a news report written in such a way that the folks back home (our families, especially) would not be alarmed at the turn of events.

The radio schedule went as planned and we sat around and yarned. Allan started his poem and together we created verses.

The Vanda Tractor Train
(or Six Antarctic Heroes)
"Driving out to Vanda on the tractor train,
"Six Antarctic heroes, men who knew no pain,
"Set off o'er the tide crack out on to the ice,
"Never more to turn back, regardless of the pain."

Every now and then someone would nip outside for fresh snow to melt for water in the largest pot we had and to thaw our canned beer. We also brewed hot chocolate and even cooked up custard to go with canned fruit. The heat from the stove made our little hut very comfortable. I sat alongside the stove that kept going for several hours. I think all of us smoked as well. I was getting a monstrous headache and a couple of the others were too. So I munched on soluble headache pills and handed the packet round.

Finally, I'd had enough and with my head splitting, I went outside. As I stumbled out and banged the door shut behind me, my first thought was I'd just had too much beer. One step into the frigid air, the world closed in around me and I face-planted smack into the snow. I lay there, shaking, freezing. I staggered to my feet and wobbly-kneed felt like passing out. Everything was yellow. I could not even manage a bedtime pee. I opened the door and told the guys I didn't feel good and was going to bed. My head pounded like it was going to fall off, and I was having trouble with my vision. I struggled through the tunnel into the tent and into my sleeping bag. I was gasping for breath and sweating. It felt like my heart or my lungs or something was about to burst through my chest. I'm going to die, I thought. My head, my head. I was shivering. What about Lois and the girls? I cried out to a God I had learned about at Sunday School. I thought of our wedding in the church and the three daughters we had baptised in the church. I pleaded with this God to take care of them all and started muttering the Lord's Prayer as we had been taught. Everything went black.

"C'mon P-R-O get up, get up, wake up, P-R-O, c'mon, wakeup, get up, c'mon," Bill yelled as he unzipped the sleeping bag and pushed and shoved me around to get dressed. I struggled. He yelled. Bill got me to the tunnel and pushed me through flat on my face into the snow. Someone helped me put on a parka, mitts and my balaclava. I guess I had boots on. The others were outside and at Bill's urging we limped, half running, around and around the sledge gasping fresh air into our lungs in the minus 30 something temperature, the wind whistling through any open clothing. A couple of the guys spewed their guts out, the rest of us almost. Not a pretty sight. After about 20 minutes of this we leaned on the sledge heaving.

"When Sam told us how he was feeling after hopping out for a pee, I remembered," Bill said. "The same bloody thing happened with me once before. Carbon monoxide."

VEHICLES IN CREVASSES

By Graeme Connell, NZPA special correspondent

SCOTT BASE, Today.—The former transantarctic expedition Snocat and a caterpillar D4 bulldozer broke through the snow bridges covering two crevasses high above the Wright Valley about 12.15 a.m. yesterday.

The drivers and sole occupants, Messrs Bill Lucy, Timaru, and Hugh Clark, Tuatapere, escaped from their vehicles unscathed. Both vehicles are part of the New Zealand Antarctic research programme tractor train which is hauling about five tons of fuel and supplies to New Zealand's new winter-ing-over station at Lake Vanda.

Plans were being made today to get the two vehicles out of the crevasses. The work is expected to take several days.

The two vehicles had been taken from the tractor train temporary camp to the start of the proposed access routes to the valley. Mr Clark was following the Snocat when the rear of his eight-ton tractor slipped through the covering.

Mr Lucy had reached their destination. He turned back to give assistance but 50 yards from the bulldozer the rear pontoons of his Snocat broke through another crevasse.

These incidents were a bitter disappointment to the six men in the party who had brought the train 90 miles from Scott Base across the sea ice of McMurdo Sound and up the Wilson Piedmont Glacier without trouble.

We'd been so busy having a warm and relaxed time in the wannigan we'd not given a thought that by thawing snow in the big pot on the small stove burner we did not have complete combustion. There was no ventilation and the hut slowly and silently filled with the unseen poisonous gas. Sam and I had been the closest to the stove.

With plenty of fresh air in our lungs, we headed to our sleeping bags and mumbled our way to sleep.

The Vanda Tractor Train heroes rest in their efforts to free their vehicles from crevasses on the neve overlooking the Wright Dry Valley. From left: Noel Wilson, Bill Lucy, Hugh Clarke, Allan Guard and Sam Magee.

DAY 7

It was definitely not a good morning when Bill roused us to say a helicopter was coming in. Grrr. My head was still not part of me but at least the pounding had gone down a notch or two. With leaden limbs I moaned and complained and barely made it out of the tent with clothes on when the chopper landed. We were all in much the same physical condition. Noel clambered on board and the chopper made a trip up the hill and unloaded the gear we had called in. Before it came back for the rest of us, we had time for a quick coffee. We loaded more gear into the chopper and hitched a very thankful ride up the hill. It was my first ride in a helicopter, but I was not in a very good shape mentally or physically to enjoy it. The helicopter crew wished us well and left us high on the neve surrounded by two stricken vehicles and all the stuff we might need to extricate them. We had little strength to do much but we did take immediate advantage of the fresh bread and butter that had been sent up by a very thoughtful group back at home base, a totally appreciated change from canned stew and soup.

Not one of us had much in the way of energy, and we worked at about half normal. We did feel a bit better after our stomach pick-me-up in the thankful sunshine, so we set to carving a trench about 20 feet behind the Sno-Cat. The helicopter had brought some wooden 10 inch x 4 inch x 10 foot beams to

Noel Wilson clears ice and snow from under our stricken Sno-Cat so that the engineers can get at the broken differential. Noel is roped and working from the crevasse which yawns beneath him.

use as a deadman to bury in a trench we'd cut at right angles to the vehicle. I now figured out what the guys had been talking about and sketching last night. I was fascinated with their strategy to dig a hole into the glacier, attach wire ropes and embed the log in the hole. The wire ropes led back to and were attached to the Sno-Cat's rear pontoon tracks that would then, powered by the engine, act as a winch. It took a lot of effort with a chainsaw, ice axes and shovels to dig the trench. It had to be deep enough for full resistance for the log as it jammed up against the face of the trench as the Sno-Cat started winching its way out. Our mechanics fired up the engine on the big orange beast and after letting it warm up a bit, Bill dropped it into reverse gear to slowly wind itself out. Just when it looked like it was coming free the rear differential broke and everything stopped once again. The unprintable words of some and the utter silence of disbelief in others expressed our bitter disappointment.

While Bill, Sam and Allan faced into that new mechanical issue, Hugh, Noel and I walked over to the D4 to start cutting the ditch for a deadman. This was a repeat of the process for the Sno-Cat.

The crevassed Sno-Cat high above the Wilson Piedmont Glacier.

The first trench we started digging turned out to be a snowbridge over a monstrous crevasse.

"Bloody hell," we chorused. Our probing earlier had missed that beauty. Fortunately we were not on the section when it broke through. Fascinated by the size, we dropped a can down but could not hear it hit bottom. With heightened senses we became more vigilant to our surroundings. Noel set about probing further to make sure we were on stable ground. We started on a new hole, using ice axes to break the ice and shovels to scoop it out. The chainsaw was now way too blunt. And the spinning blade only served to melt the ice. Digging required a great deal of effort and it took all our strength to just lift a shovel towards the end. My thoughts by this time were about sitting on a surf beach back home sunbathing. I chattered on telling the bachelors how nice it would be to be at home with Lois. I think they wanted to drop me down the crevasse.

With the new trench completed, we hitched the wires to the grouser plates on the D4 tracks. Hugh had a shot at starting his machine. Not a

Preparing to cut and dig a trench for a deadman anchor to enable the Sno-Cat to winch itself out. The D4 can be seen in a second crevasse.

show. The diesel engine, as expected, was frozen solid. We packed up our things and headed down the hill. It was mid afternoon. We were truly beat and didn't say much. Noel and Hugh hustled up a brew of tea, and we sat outside in the lee of a sledge enjoying the afternoon sun. We did not want a repeat performance of the night before. We yarned after our early supper of stew and plastic spuds and for sweet relief, continued adding to the poem as Allan put it all together.

"The day dawned bright as

work began to drive the Sno-Cat through,

"But as the last pontoon was clear the crown wheel split in two,

"With diffy smashed but Sno-Cat safe our heros turned for bed,

"And soon they had an order on the next sched."

Our scheduled midnight report to Scott Base went well and we hit the sleeping bags.

DAY 8

With stricken vehicles, we were still a long way from the objective. Our new day followed the usual pattern of get up, feed on whatever looked good and make the climb up our mountain. Our first task was to get the D4 out of its slot using its own power. But we had to get the engine running first. Hugh and Sam wrapped the engine in canvas and using a petrol-fuelled heater that had been flown in the day before blew hot air around the engine for a couple of hours to warm it up. After a couple of attempts Hugh finally got the big motor going. It was left idling to fully warm up. Hugh took the rear window out of the cab and attached long ropes to the control levers inside to avoid anyone having to be in the cab because of the danger of it tipping back and down into the crevasse. Thankfully the plan to use the deadman-wire rope-grouser plate winch worked and the D4 bravely churned its way to solid ground with Hugh and Allan "driving" her from behind using their reins as if they were walking behind a horse-drawn plough.

We cheered and whooped, whooped around. We had a victory, one vehicle out and operating. Bill and Noel probed and marked a crevasse-free trail to the rim of the valley and Hugh drove across without incident. Attention now turned to the Sno-Cat. The geniuses figured out a slightly altered system now we had the D4 to safety. Using

Hugh Clarke checks the fuel tank on his crevassed D4 Caterpillar.

a pulley system from the D4 to the Sno-Cat to the deadman they figured the Cat would be able to pull the undriveable Able to level ground. We held our breath as Hugh slowly and evenly applied tension to the wire rope. Easy does it, we breathed, as he took up the slack on the ropes. Then bang, the ropes went slack again as the deadman flew out of the trench. It sure looked like the system would work. We just had to dig a deeper hole. The next attempt held and we finally had the big veteran on even ground.

It was time to call it a day and we trooped downhill once more to our friendly little camp and checked out the gear the chopper had brought in earlier in the day. There were replacement parts for the Sno-Cat and fuel for the new stove delivered the day before. We did not want a repeat performance of our carbon monoxide night.

DAY 9

It was a late getup again, and by the time we'd breakfasted, it was past noon when we set off on the climb. The day's objective was to get the Sno-Cat to safety. Our efforts the day before had resulted in the vehicle being just clear of the crevasse making it possible for Sam and Allan to repair the broken differential without toppling down the yawning crevasse. I really admired those guys scrabbling around on the frozen surface in bulky clothes working ever so carefully with ice-cold tools and bare hands when necessary to adjust and fit. That job done, we heard Kiwi adjectives that would silence the noisiest bar back home when Sam and Allan discovered a shattered universal joint. What more could go wrong with this vehicle? And to add insult to injury, the on-board spare part was too big!

Bill and Allan put the breakages and failures down to the age of Able as well as general fatigue in the extreme low temperatures. So far nothing had happened that could not be fixed. The biggest problem was that we were so far away from equipment and spare parts. Able was big and strong, a power horse that had completed the first-ever 2000 mile motorised crossing of Antarctica some 12 years earlier.

Allan Guard attaches wire ropes to the D4 while warm air is pumped around the engine from a portable gasoline powered heater.

It wasn't so much the breakdowns that prompted the air of despair. "We've been at this too bloody long," a dejected Bill said. "We're running out of time."

To complete the traverse, the supplies had to be dropped into the valley and the vehicles had to high tail it back to Scott Base so further supply runs could be completed before the sea ice became too unstable for travel. Dedicated helicopters in this year of economic restraint were out of the question. Our work had to be done, the Vanda Station had to be built and ready for a team to winter over. We got a bit grumbly and voiced mutinous thoughts about the nutters back in the bureaucracy and questioned the sanity of such a method to establish Vanda Station, an important new scientific venture for li'l ol' New Zealand.

Our venting done, Hugh and I left the group working on the Sno-Cat and took the D4 to see what could be done to get a route down into the valley. The D4 struck impossible ice and Hugh turned back. Noel and Hugh went off with the D4 to have a crack at negotiating the valley down the moraine slope. This proved impossible too as the 'dozer could not move even

The D4 has made it to safety and the Sno-Cat is free from its crevasse.

the smallest rock from the solid frozen talus (sloping mass of rocky fragments at the base of a cliff). We abandoned the evening radio schedule and worked on. My fingers and toes were blocks of ice. I pitied the guys working on the vehicle: if I was this cold what about them? We'd lost any warmth from the sun and were now in cold shadow. The never-ending wind was vicious on this exposed mountainside. When partial repairs were completed on the Sno-Cat, Bill fired her up, loaded the gear on and tried to move her across the glacier just using the front pontoon drives. No go. We took gear off to lighten her up, but that didn't make a bit of difference. Able was not able. Little was said save for Bill's "To hell with it, let's go home." What's more, the rocky ice-packed terrain was proving too much for the D4. Down at camp, one of the tractors had busted a track earlier when we were shuffling the sledges. We now had three vehicles stranded!

After the usual potluck cookup we hit the sleeping bags. Hugh proved to be our comedy act that night. He had a way of telling a story that made us all laugh. I fell asleep feeling much better about the events of the day.

DAY 10

Friday started off on a sour note as we could not raise Scott Base on the 8 am schedule. We breakfasted and around noon headed out in different directions to determine the next course of action. Sam and Allan set off on the remaining Ferguson tractor to look for an alternate route further along the Wright Lower Glacier. Noel headed off to the far side of the glacier to see what might be possible and Bill went to take a close look at the bluff.

Hugh and I manhandled two five gallons drums of fuel up the stiff mile climb to the D4. Our outdoor lifestyle, surviving in the bitter cold and climbing up and down this mountainside every day had lifted our physical fitness to a new level. Hauling the drums was tough going, but we made it and emptied the fuel into the D4 tank. The Cat had been left idling through the night but when Hugh gave her the throttle, the engine died and nothing he did would bring it back to life. We could just make out the others way below us, Noel a pinprick on the horizon, Bill making his way round the bluff, and the Fergie seemed to be making some progress. We met up with Bill and told him about the split fuel line we'd found on the D4. It was out of action till the line could be repaired. We'd have to see what Allan and Sam thought about that.

Together, we took a long walk around the moraine, up and down the glacier to the valley floor. If only we could get off the neve and over the ice ridge. The "natural" path we were on was the only one that seemed feasible to get the supplies into the valley for later transport by wheeled vehicles to the Lake Vanda site. Sam and Allan joined us and we plodded on into the valley and enjoyed the magnificence. We got used to the non-stop wind, reportedly at almost 30 miles-an-hour all morning to the east and then 30 miles-an-hour to the west in the afternoon. These are katabatic winds that sweep down off the polar plateau. I had to be extremely careful with the cameras as the wind carried with it fine sand; any exposure to the elements would be akin to sandblasting. I'd been warned that the sand would find its way into the lens barrel.

We inspected a couple more mummified seals and stopped in the lee of a rocky outcrop and enjoyed a cigarette. Down but not out, we headed wearily back to camp. No progress today. The challenges seemed too enormous to overcome, and although I did not voice it, I worried about my real job, the things that were not being done, the stories that were not being written. It was satisfying being out in the field and in such good company. I am sure I was a useful part in the team but I had to remember why I was in Antarctica.

I anticipated a growly memo from Antarctic Division about output as I had been reminded rather bluntly before coming to the ice that my function was to keep New Zealand's endeavours in the public eye. That meant keeping the media well supplied with upbeat material. I had to be cognisant of the political overtones.

I prepared a news item to update progress for the New Zealand media, saying the vehicles were out of the crevasses but out of action and that the team continued to seek the best route in trying circumstances into the valley. We did not have any luck with the radio that night. We sat around and yarned a bit to get our spirits up.

Talk was different tonight as Bill promoted the thought about our Antarctic heros. He mentioned Borchgrevink, the leader of the first team to winter over in Antarctica, up at Cape Adare at the turn of the century. We chatted about Shackleton and his journey south and how he had the guts to turn round even though he was in range of the South Pole. We reflected on the time of the year and how 53 years earlier his ship Endurance was jammed in the ice in the Weddell Sea, on the South America side of Antarctic. We talked about Scott's northern party in 1912 fighting for their lives, stranded and marooned in horrific circumstances for a long winter, not knowing that Scott and his Pole party had succumbed in a blizzard just 11 miles short of One Ton Depot. Our troubles paled in contrast as we recalled polar history and hardship. We finally hit the tent about 11pm. Hugh kept us in fits of laughter with his stories, and again I fell asleep feeling pretty darned good about things.

DAY 11

I was half asleep when Allan casually announced there was a chopper coming. He could hear it. The rest of us couldn't. Seconds later Bill stuck his head in the tent yelling us to get up and get out el pronto as the unscheduled chopper was already circling the camp and coming in for a landing.

"P-R-O, we have a VIP with us, Captain Charles Upham VC, " Bill announced.

Total panic. I exploded out of my sleeping bag. Here was story and

More work on the D4 repairing a broken fuel line.

picture falling into my lap and I was totally unprepared. We all were. But this was one for the media. I fumbled around. It is not easy dressing in a hurry in a tent with three other guys. I fell out of the tunnel virtually at the feet of our famous guests and ran to the wannigan to get a camera. But I loaded the camera too fast and it frosted up. Darn, darn, darn.

I blew this one. We were all scrabbling around when Captain Upham's hosts, the US Navy's McMurdo boss Admiral Abbott and entourage gathered with us in front of our tent for a photograph. Captain Upham and his friend Clutha MacKenzie were very amused at our isolated camp high up on the neve and got a laugh out of our little pile of empty beer cans alongside the wannigan. It was a flying visit for this famous Kiwi, one of only three people in the world to be awarded the Victoria Cross twice for bravery during the Second World War.

The group was only on the ground for minutes, enough time to say hello, drop off some fresh supplies, and snap a photograph before the chopper whisked them away in the doubtful radio conditions. The whole visit was over almost before it started. I would have liked to had a newsy chat with our visitors to send out with a picture if and when I returned to Scott Base. A few days later my US Navy opposite gave us all a print of the picture he took in front of the tent. The picture shows me kneeling in front of the group with a camera, Noel is still getting out of the tent and Sam did not make it.

Once we'd gathered our senses after this US Navy photo-op, I cooked up

A US sponsored whistle stop by a famous New Zealander Captain Charles Upham (left). Fourth from left is Bill Lucy, Lt Dan Davidson, US Navy , Hugh Clarke, Clutha MacKenzie, Noel Wilson, in front, Graeme Connell and Allan Guard.

a snappy breakfast of scrambled eggs and bacon. Mmmm, did it smell good, and it did wonders to our morale. We split into two parties for the afternoon – Hugh, Allan and I headed to the D4 and the others headed off to repair the track on the Ferguson tractor. Using a small gasoline generator (destined for Vanda) and an electric iron, Allan soldered the broken fuel pipe. It was a painstaking operation but Allan's creativity solved what seemed like the impossible. Cranking the engine, recharging the batteries with the generator and using lots of ether spray, we finally awoke the 'dozer from its frozen state. We left her idling happily and, frozen to the bone, made our way back down to camp and a great meal of stew, fried-up vegetables and plastic spud. For desert, Sam went creative and made up an Antarctic special, including coffee, honey, trail biscuits, milk powder and sugar. Ok to a point, but a bit gluey and chewy. That night we sang and chattered in our tent for some time. We finished the day at about -25°C, comparatively warm, I thought.

DAY 12

There didn't seem to be much point in getting up early on Sunday morning. We'd done all we could do and we were waiting for parts for the Sno-Cat. It was an ideal time for us to update our notebooks as we waited for the midday radio schedule. Conditions were good and it was great to hear Robin tell us a chopper was on its way with 1000 pounds of supplies for Vanda as well as replacement parts for us. It was remarkable how Robin was able to get our spare parts and equipment flown in at such short notice. The flights were part of the allocated hours in the US-NZ agreement, but these unscheduled trips cut into the allocation. I think also Robin's good relations with the guys over the hill blurred the lines a bit, and the USARP guys added a drop to us while on a mission of their own.

"Better send P-R-O in," Robin advised on the radio call. "I'll manifest him on the return trip today."

This sounded like the right move to me. Apart from extra hands, there was little I could add to the operation. I felt a bit awkward at baling out as I gathered my things together. We saw the choppers fly up the Wright Valley to Vanda about 3:30 pm and an hour later they stopped by to unload at Camp 4. It was a fast turnaround. No time for fancy goodbyes. I clambered aboard, the door slammed and we were up and away. As the wheels cleared the glacier I saw Allan rush from the wannigan waving what looked like mail. Too late, we were in a hurry, the weather was deteriorating rapidly, and the pilot did not have radio contact.

"The sabbath dawned all overcast, the wind was icy chill,

"Two choppers flew in with the mail and spares for up the hill,

"They took P-R-O away, it was a sad'ning sight,

"But the five South Islanders would carry on the fight."

I'd have loved to have those letters from Lois. Now I was triply sad, leaving the letters, leaving the tractor train and above all the unfinished story.

What did Lois have to say? I'd call home tonight if there was space for me on the tight evening schedule. I smiled at the thought of my friends examining Lois' envelopes. She draws hearts on the flap. That'll give them a charge, I smiled, wondering too if these ones were perfumed.

Meanwhile, I was in for a rough trip back to Scott. Our chopper was starting to dance around a bit as the wind picked up speed. The loadmaster looked out the window not wanting to make eye contact with his passengers. I felt my survival bag. Yep, it was right next to me.

We were now flying over the sea ice, heading towards the seal hut and the place of our first camp. My second and longest trip in a helicopter was turning into a memorable ride. The machine bucketed about, shaking, bouncing up and down and slewing from side to side.

Hell, I'd heard about these things shaking to bits when they came in to land. This had something to do with vibration and the effect of rotor speed and updrafts from the ground. I didn't want to know the technical explanation. This was a white-knuckle flight.

Conditions worsened rapidly out over the sea ice and the pilot indicated he'd have to get out of the sky and was heading directly to the helo pad at McMurdo. He would not take the chance to land and drop me off at Scott Base. The wind was now head on at around 35 miles an hour, moderate gale force when trees would be waving around back home and it'd be tough to walk upright.

I reminded myself of the simple fact that these guys all wanted to get home as well. We made it rough but fine into the chopper base at McMurdo. The flight crew was welcomed with cheers and shouts from their ground-based mates. Whew, I thought, it must have been tougher than I figured. They were a hospitable lot and took me into their flight headquarters and plied me with fresh coffee as I waited for a road pickup from Scott Base, just four miles away over the hill.

Geoff, our hard-talking New Zealand navy cook, had volunteered to come pick me up. Skua gulls were on their eternal search for food and seals lolled on the ice as we came over the hill and looked down on the green huts of the Kiwi base. There appeared to have been a bit of a thaw.

A grubby, smelly old Antarctic explorer, I'd earned my badge and got a great welcome from the guys who all wanted to hear the tales of hardship. With very little effort, I conned someone into giving up their shower and laundry night and took on the House Mouse duties for the next day. I totally enjoyed the luxury of my brief shower and even gave a chuckle as I stuffed my clothes man-fashion all at once through the washer and dryer. New clothes, good food, and friends did little to take away my sadness for the guys up on the neve struggling against the whole of Antarctica to get materials, fuel and supplies in place for the new Vanda station. I stayed around the radio room for the midnight sched for one more quick chat and to hear progress.

Oh, what glorious luxury to wrap myself in clean pyjamas and crawl into clean sheets on a soft mattress and to put my head on a pillow.

November 4

I bounced out early at 6:30 am to get a tractor train progress report out on the morning schedule. New routines took over. It was payback time for the luxury of water.

I called Lois that night. She sounded very sad and all I could promise her was a good long letter. "You did not sound as chirpy as you usually do," I wrote in my letter between mouse duties. To make the letter newsy for her I wrote it as the day unfolded. A piece here and a piece there. After a couple of pages it was time to stop and prepare for breakfast and wake everybody. By 8:45 am, I was snuggled into my ski pyjamas again and in bed adding a few more lines before grabbing a couple of hours sleep after a long night on watch.

"During my watch," I wrote, "I shaved round the edges of my ruddy red beard. The sides looked a bit scraggly but the mo' and chin are beaut." I fell asleep writing the letter. I still had a day or two before the next mail went out.

November 5

"We had a rude awakening this morning at 6:15 am," I wrote. "The fire alarm went off in our hut when one of the heater units went berko. We were all up and running, fire extinguishers in hand. But it was nothing. That is the second alarm we have had now. Fortunately, no fire. That is our greatest danger here.

"It's so very dry. Nostrils dry up, but I'm sure I'll get used to it soon. I didn't notice it so much out on the trip. And I had a haircut last night. Keith, in the lab, did it. Not a bad job, either.

"Right now I'm busy preparing programs for some visitors we have coming and I want to try and sneak a picture of Robin for the notice board. Its his birthday and, therefore, his shout tonight."

I ended that letter with the thought that "from what has happened to me since I left, I think I will be a better more loving man for you..."

November 6

After dealing with school mail, getting a parcel and letter on the plane for Lois and the girls, letters out to Wellington for more book supplies, brochures and pamphlets I buried my head in the story I'd written about the tractor train. Overall, I was not happy with the results, but it was a satisfactory story. I was bushed from the day's efforts, largely from being inside all day after such bracing outdoorsy time in the field.

It took almost another week for my friends to achieve the objective of getting the supplies and materials to the valley floor for a group including Hugh and Noel to ferry to the Vanda Station site with wheeled vehicles. Hard to believe, but Allan and Bill made it back from Camp 4 to Scott Base with the Sno-Cat and the bulldozer in an 18-hour non-stop dash.

"When it became evident that we could not make the track down through the boulders, we went a bit further down the neve and tried to make a road off the glacier by pushing snow down hill into the valley to make a snow road where our Diff Glacier and the Piedmont meet," Allan explained. "The slope proved too icy for the D4 to get any grip, and it could not reverse up under its own steam, so we towed it back up with the Sno-Cat which has sharp little grouser teeth to give it grip on the ice."

Once the D4 was safely back at Camp 4, the guys explored further north along the terminal face of the Wright Lower Glacier and located a place where the tractors and equipment could be lowered into the valley. They used the Sno-Cat with 400 feet of wire rope flown in after being recovered from the Scott Base dog lines. "The tractors were pushed over the edge one by one, tethered to the Sno-Cat which drove slowly forward lowering them down the cliff," Allan reported. "The trailers and pairs of fuel drums were lowered over in the same manner. One of the loaded trailers got away on us while going over, and spilled its load all over the place, but apart from a dented drum or two, nothing was lost."

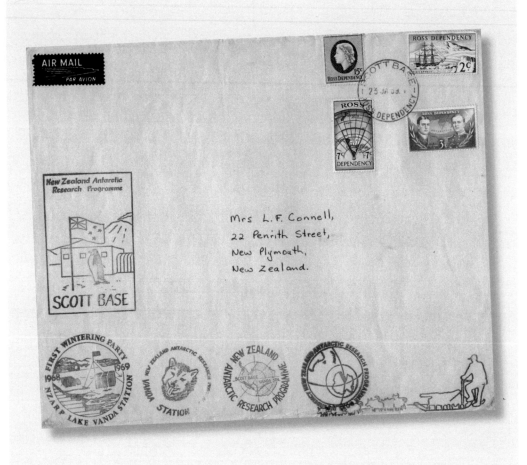

6.
LETTERS FROM THE
BOTTOM OF THE WORLD

Early November

Hello My Darling:

Golly, I was a bit sad when I went to bed last night after talking with you on the phone. You did not sound as chirpy as you did last time I rang. But then, no wonder you've got a helluva job to do there on your own. And I am not much help way down here. I love you my dear. Please don't fret. Each night I go to bed thinking of you. The other night I thought you were right here beside me when I woke up. I went to put my arm around you but you were not there.

I'm on late night watch which means going to bed till 1:15am then up till 8:15am and if you are lucky back to bed till 12:15pm. It is just 6.30am now. After doing my chores I spent a few hours in the darkroom. Some of the prints I have sent you. The enlarger is not in good condition. It's an Omega, good quality but roughly used in the past. Most of the time I spent trying to proof my field photos. Buggered if I can do it. So now I am going to look up a book we have here. I have also sent up a box of slides. After they have done the rounds please send them back as they have to go to Antarctic Division to see if they want copies.

'I had to stop there to get things ready for breakfast then wake the whole mob for breaky. It is now 8.45am and I am sitting in bed with my ski jammies on (nice and warm) putting down a few more lines before grabbing some sleep. I feel pretty weary. As I explained last night (on the phone) I arrived back here on Sunday night by chopper. As I was very dirty, very smelly, very tired (my last shower was about October 15) I asked if I could have one. The downside to this is to swap with a guy on the house mouse roster. I had to sew some fly buttons back on my trousers this morning. The cotton is not much good with these issue trousers. I used 3oz thread. Its like string but I know they won't fly off again. It is not an area I want to get frost bitten!

Well, I fell asleep writing this so now another day has passed. Next mail out is tomorrow I think.

I did check out that money issue with Antarctic Division. Apparently you will not start to receive the Antarctic allowance for about a month. That is paid from the day I left New Zealand. Thank you my sweet for the big job you are doing looking after our darling little girls and our other New Plymouth interests.

PS. Final thought ...would you send me some halibut oil pills as they may help with the dryness.

Your ice-bound lover

G

October 1968

Hello Sweetheart

Lance and Ally stayed for supper. All full and content, Ally got into the bathtub with Bridget while Lance took the girls down to the playground. They got back in time and the girls jumped into the tub with Ally and Bridget. Then they all had fun washing Ally's really, really long hair. It was a lovely scene, full of fun and laughter.

Dear ol' puppy seems to really miss you. She sits up at the gate waiting, hopefully. She seems to be going through an eat-everything-in-sight stage, even to nipping when she's playing with me and the girls.

Hilary told Mother today that you were going to call us when you get back from your trip. I hope so too!

After reading your letter, I feel a bit lousy telling you when I feel sad. In fact, yesterday I had a meltdown as things went all wrong. To cap it all, I've lost one of my rings, the one a wee bit wider than my wedding ring. I had my rubber gloves on and pulled them off and off came the ring. I did not notice until I was down fixing the lawnmower as the handle had come off. Sorry, my sweet. I should be more careful.

Some of those rocks you have sent still feel very cold when held in the palm of my hand. I like the pinky ones. They look so ancient.

Don't worry about the money. Last Wednesday I got $117, and I have even managed a payment on the car repairs; car and house insurances are paid and the phone bill of $12 with our calls from Wellington and Christchurch, the power bill and even the mortgage. So I haven't been able to save much yet!!

Till next time my darling one,

Tuppy

November 13, 1968

Dearest

Things have been hectic around here. I've been working from the time I get out of bed at 7:15 am till around midnight. Perhaps I am trying to do more than expected, but I feel it is all worthwhile. Tourists are starting to bug me as there seems to be a never-ending stream of them.

This is a rough message and I am writing it between getting photos off to newspapers and NZPA. To do pix for NZPA, I have to run 20 copies of each print so trying to fit that in to my day can be tough. I just hope they are o.k. So far I have not had a chance to do any Antarctic pictures for myself.

Tonight's plane is the first out for about five days and the stuff I have ready to go is from yesterday and today. I called my Dad today for his birthday and I had hoped you'd be there so I could get two for the price of one, a gidday to Pop and long chat with you.

I met the US chaplain today. He came over for a visit and invited us to join him on Sunday at the Chapel of the Snows. I like him.

Nice writing love letters, isn't it?

Scott Base scribe

G

Lois always decorated her envelope backs.

Rachel's first attempt at a long letter.

Technician Keith Mandeno tends his flower garden in the Scott Base lab.

7.
NEWS TO MAKE NEWS

"Richard Nixon has been elected to succeed President Johnson".

This was the succinct first line of the first Scott Base Newsletter I introduced to our news starved lives. I started the newsletter November 7 in the hopes that it would become some sort of a reciprocal (international and New Zealand news in exchange for their NZARP news) self-generating information sheet to help me keep abreast of everything going on around our summer activities and in turn allow me to get news, pictures and feature material to the New Zealand media.

The first sheet of the two-pager contained international and New Zealand news items gleaned from radio news by our Post Office techies. The second page contained all the stuff that was happening in and around our little world at Scott Base: weather, next mail out (and in), comings and goings, and what the field parties were up to. I also posted information on when I sent out a news article so that the fellows could tell their folks at home to watch their newspapers for news. My goal was to post it on the bulletin board in the mess hut before lunch. Throughout each day, I'd note snippets of info and followup with key contacts to complete the item. After the Post Office blokes gave me their stuff in the morning, I'd two-finger type the newsletter with carbon copies on my trusty typewriter.

On the newsletter's launch day, my friends on the tractor train reported progress in getting the D4 off the frozen rock after resorting to hacking a route with ice axes. The following day, they abandoned plans to cut an access road into the valley, opting instead to lower the gear and the two tractors over the glacier about two miles north from our Camp 4.

Lab technician Keith Mandeno gave me a story when he began growing plants in the laboratory. Keith had asked his mother in Auckland, New Zealand, to send some something down for him to try. The result was six potted plants, some from the home garden and some from an Auckland plant merchant who "guaranteed to send plants anywhere." When he was told to ship them to Scott Base, Antarctica, he figured it was a joke. Finally convinced the request was on the level, the merchant shipped two Lilies of the Valley, a rose and some variegated ivy free of charge.

[In an email February 2010, Keith wrote:

The Lily of the Valley flowered and then died down for winter. Growth returned before I left the ice but the plants were much smaller.

The ivy died with undue haste, as I recall.

The rose - well that's strange because I have absolutely no recall!

There were some seeds as well and these germinated but did not enjoy the artificial fluorescent light much.

As I recall the next incumbent was not interested and so the lot went into the bin during my last minute tidy up. It's possible I brought the Lily of the Valley bulbs back to NZ.]

More nuggets:

[Thursday] November 7

Canterbury University zoologist Geoff Tunnicliffe arrived last night for a busy month collecting specimens for Canterbury Museum's Hall of Antarctica for the museum's centenary in 1970. As soon as the weather improves, he will high-tail it to Cape Bird until around November 13 and work with two of the university's biologists Eric Spurr and Morgan Williams. Geoff will collect Adelie and Emperor penguins that will be sent back and prepared at the museum. He also wants to replace and enlarge upon some of the museum's vertebrate specimens that originated with Scott's 1910-12 expedition. He plans to get up to Cape Hallett too for snow petrel and Weddell seal specimens for both display and scientific purposes. And in between all this he will photograph birds and animals in their natural environment to assist in setting up the museum displays.

Visitors today were Dr Phil Sulzberger, Deputy Director of the Australian National Antarctic Research Expedition and two Americans, one of whom has completed a 12 month term at Byrd Station and the other as head of a US polar group.

[Friday] November 8

Scott Base received its largest airlift of supplies. About six tonnes of scientific equipment and food supplies were unloaded from a Super Constellation. Much of the equipment was for the new wintering over station at Lake Vanda. Scott Base's main supply will come with HMNZS Endeavour in January and February.

[Saturday] November 9

Post Office technician David Blackbourn received $US4000 worth of new gear to replace two of the Phillips field receivers. The two new Collins sets (the best around in New Zealand) were designed for use with the Labgear field sets.

The Tractor Train crew reported success in their efforts to lower the Fergusons over the Wright Glacier.

Chippy John Newman had to get up at 4 am to finish work on the new phone booth in time for Blackbourn to install the phone for the 1:30pm schedule.

Wayne Maguiness, the base mechanic, was late mouse but decided he could not spare the time to go to bed that morning as he had urgent work to do on the Ferguson 20 he's converting for use in an ice drilling program to determine heat flow through the bottom muds of McMurdo Sound.

Sunday trips require Robin's approval, any trip has be advertised for all-comers and the trip book has to be filled in. Of note also was a vehicle heading for the Chapel of the Snows across the hill for Sunday service and a growler that mailbags, as the property of New Zealand Post, could only be opened by the base Postmaster or his assistant.

[Monday] November 11

New Zealand Antarctic Society visitors Frank Gurney and Jack Folwell had their stay extended by a day or two because their aircraft had to be used for planned photo reconnaissance work. This was the domino effect of a C130 Hercules crashing a couple of days earlier at Brockton Station (a US weather station on the Ross Ice Shelf). No one was hurt in that and repairs were made on site.

On the social side, base electrician Chris Rickards offered rides with the dogs on their exercise runs while mechanic Wayne Maguiness wants to set up a ski party out at the Scott Base ski hill.

[Wednesday] November 13

The sea ice is starting to break out about a mile north of Cape Bird, reports Morgan Williams who returned from three weeks in the field. The fast ice has resulted in the Adelie penguin residents coming in to their rookery much slower this year because of the distance they have to walk. Normally the population is around 50,000, but it has shrunk to 30,000 this year. Also, about 20 Emperor penguins daily have been spotted compared with two or three a day this time last year. Morgan estimates about half the skua gull population has returned to their Cape Bird territory this year. They've been arriving rapidly in the past two or three days. Breeding is well underway. Another point of interest is that the female relies on her mate for food throughout the breeding season. The females started begging as soon as they arrived.

Other Canterbury University personnel, Dr Euan Young, Tony Harrison and Trevor Crosby left for their Cape Bird location to join Eric Spurr. They'll be there till early February.

[Thursday] November 14

We had five interesting visitors this morning. The party included two Englishmen and three Norwegians. One of the UK visitors was Ken Blaiklock, who was on his first trip back to Antarctica since his involvement in the Fuchs TransAntarctic expedition a decade ago. Also visiting today were three Norwegians who will go work in the Kraul Mountains (opposite coast to McMurdo in the Queen Maud Land) area. This is Norway's first visit to Antarctica since 1959.

Ian Stirling and his seal mate David Greenwood took off this morning on a seal census in the Cape Hallett region north of Scott Base. This work is being done in conjunction with a US reconnaissance flight.

Four men were ready and waiting for a helicopter flight this morning to Hogback Hill but the chopper forgot them. Bill Lucy, Nigel Millar, John Newman and Bruce Brookes had to wait until one of the two choppers assigned for the airlift of the old Cape Royds hut returned and refueled to take them out to assist with the loading. The choppers were called in to take the hut sections to Vanda where Noel Wilson, Derek Cordes, Hugh Clarke and Alan Magee were waiting to unload. Each chopper was expected to make 3-4 trips. The hut was dumped last summer at Hogback Hill at the south fringe of the Wilson Piedmont Glacier. The hut was originally scheduled to be uplifted by the tractor train but the earlier difficulties encountered prevented this.

A Hut's standard of cleanliness is dropping. The floors are being done but other things such as window ledges and bookshelves are being overlooked. Also, too many people are leaving their late night dishes scattered around for the late mouse to pick up. The thought for the day is: He who makes muck, cleans muck.

Arrangements are being made to supply a set of Ferguson tracks to the folks at V X-6 squadron. These will be fitted to a tractor for use in hauling the helicopters around. So, the trade is not all one way.

After facing the rigours of the Wilson Piedmont Glacier, Base Engineer Allan Guard injured his right leg yesterday while on a work party gathering ice for the melters. A check with the doctor at McMurdo today revealed a torn cartilage. Allan slipped down a crevasse about 15 feet deep but just to his armpits. David Blackbourn also had an experience with a "slot". Footnote: Allan was talking with an American at the time he subsided.

Four Italians are due tomorrow for an indefinite stay. The leader is Carlo Mauri who visited last year. With him are Alessio Ollier, cabinetmaker, alpinist, and guide (he made the first ascent up a tricky face on Europe's Mont Blanc); Ignazio Piussi, carpenter, Italian ski champion, member of the world champion bobsleigh team; and Marcello Manzoni, geologist.

Technician-in-charge Peter Lennard.

[Friday] November 15

Peter Lennard, technician-in-charge, sent the following statement to the Scott Base Newsletter for publication today.

All staff, particularly housemice, are requested to resist the temptation to touch, fiddle with, open, look at (or any other verb) any item, equipment (or other noun), in the laboratory. In particular, closed darkrooms must not be entered, record tubes must not be opened, and inviting knobs on equipment must not be turned. "Das rubbnecken sightseeing, housen-moose dummkopfen keepen mittens in das pockets, relaxen, und watch das blinkenlights".

The wind the Wright Valley was blowing at around 30 knots this morning and visibility was low. Yesterday two choppers made three trips each with hut sections from the Hogback Hill pickup to Vanda Station. There are still two trips to complete the airlift.

At 9 am tomorrow our first party of visitors, eight journalists, are due from McMurdo and will visit Scott Base until 10:30 am. From 1:30 pm to 3 pm we'll host a group of foreign exchange representatives from the countries that are signatories to the Antarctic Treaty. A general tidy up is requested.

[Saturday] November 16

Deputy Leader Bill Lucy is keen to have all the drift cleared from around the huts. If each man used a shovel during an odd moment during the afternoon it would prevent evening work parties. The drift has to be cleared to a suitable distance to be picked up by the bulldozer. The sooner this work is done the better.

Glaciologist Arnold Heine and surveyor Alistair (Taffy) Ayres were due back today after three days on the ice shelf. Taffy should now be well initiated as an intrepid shelf man.

For the record: Thirty-six visitors to Scott Base since Tuesday have signed the Visitors' Book. A busy week.

[Monday] November 18

Two weeks ago CBS cameraman Peter Good, was in Nairobi. Today he completed shooting the Scott Base huskies with Chris Rickards to include in a 30-minute documentary he is making on animal life in Antarctica. The film will be shown over the CBS network sometime in February or March. The dog piece is expected to make up about a third of the programme. On Wednesday, Peter flies to New Zealand where he will spend a month filming NZ wildlife for the same programme. The work is being done in conjunction with the New Zealand Tourist and Publicity Department and Air New Zealand.

Italian expedition members Ignazio Piussi and Marcello Manzoni catch snow on their geology tour through the TransAntarctic Mountains. (Marcello Manzoni photo)

Italians Marcello Manzoni and Ignazio Piussi will head to Vanda for an indefinite period. The other two Italians, Carlo Mauri and Allessio Ollier will accompany the VUWAE 13 group to the Boomerang Range early next month.

Bruce Brookes will accompany Jim Cousins on his ice-drilling program between Dellbridge Island and Butter Point. Jim will put down a series of probes to determine heat flow through the bottom muds of McMurdo Sound. This work is expected to take about three weeks and then Jim will head to Vanda. The departure with the newly converted Ferguson 20 tractor has been held up because of a delay in the arrival of equipment from New Zealand.

[Tuesday] November 19

The largest cosmic ray disturbance since 1962 has blotted out all radio communication and grounded all USARP flights. The last major disruption occurred in January 1967 when two successive disturbances lasted two weeks. The prediction for the present blackout is that it will be at least a week before there is any improvement.

Complete radio blackout conditions began at 5 pm yesterday. The Scott Base lab staff first heard at 11 am yesterday that the US satellite Doppler program was being interfered with. A check with the Scott Base ionosonde–the only one in this part of Antarctica—showed that all the layers (both E and F) had disappeared. McMurdo's riometer hut showed that an absorption began last night at 10:21 pm. A check through the hut's records showed the disturbance to be the largest since the riometer program began in 1962.

The riometer registers the degree of absorption in the ionosphere on the 30 and 50 megacycles frequencies. Scott Base's ionosonde scans the ionosphere between 1 and 22 megacycles and measures the way in which the signal is reflected back to earth. E and F are two layers of ionized gas. The sun, breaking down molecules in the upper atmosphere, causes the blackout. The disturbance is so intense all signals are being totally absorbed.

[Thursday] November 21

Sixty American millionaires arrive in their Convair 990 at 12:30pm tomorrow. The 64 visitors will visit Scott Base about 3:45pm for a show of the dogs and sledges. These super tourists are on a world flight across both poles to commemorate the 40th anniversary of the first flight across the North Pole by Admiral R.E.Byrd. Money raised by the passengers will be used to establish a polar studies centre in Boston, Massachusetts. The flight cost each traveler $10,000. After the 30-minute Scott Base visit the plane takes off for South America via the South Pole.

Seven US Congressmen are due to visit our base on Saturday at 2 pm.

Communications disappeared soon after 4 pm today. Up till then the radiotelephone schedule had been operating quite successfully. Peter Lennard advises that an F layer and some sporadic E layer reappeared this morning. The ionospheric layers had taken up what was generally known as a G condition when the F2 layer was very high, about 800 km (normal is 350 km) above the earth. Peter says the ionosphere is still extremely disturbed and would not commit himself on any prediction for improvement.

[Saturday] November 23

Comings and goings. A helicopter flight to Lake Vanda yesterday started a big shuffle of personnel. Alan Magee has ended his sojourn in the Wright Valley and will return to New Zealand tomorrow. He is the last member of last year's team to leave the ice. John Whitehead (Victoria University field assistant) and Tony Bromley (meteorologist) will go in. Whitehead will return about December 1 and Bromley about December 20. Italians Marcello

Manzoni and Ignazio Piussi will go and remain in the Wright Valley till the end of summer. En route, they helped Lucy and Newman with the final loading of the Cape Royds Hut from Hogback Hill. Hugh Clarke returns today, Derek Cordes is expected out Monday and Chris Rickards goes in to do the electrical installation in the reassembled huts aka Vanda Station. With six men up there, things are pretty cramped. Tomorrow a Hercules will take Robin Foubister, Carlo Mauri, Kaj Westerskov (Otago University), Brian Johnston, Don Robertson and Geoff Tunnicliffe on a 400 mile flight to Cape Hallett on the northern coast. Foubister will return with the Herc while Geoff, Carlo and Kaj will be there for about a week. Brian and Don will be picked up by icebreaker at the end of summer.

The first regular weather report from Lake Vanda was received at Scott Base today. Tony Bromley has started the program that will be taken over by Ron Craig who will winter over. Observations are being made every three hours and sent back to Scott Base every six hours. Today's report was: cloud 3/8ths cover, wind WSW at 15 knots, temperature -29°C. Pressure 980.6 millibars.

Leader Robin this morning made an intensive study of the base First Aid boxes. The main thing he found missing was scissors. "Please do not remove," he asks. If during the following months supplies run short in the kits, he will see they are replenished. If any person feels there is something not in the kits there is probably a good reason for it. Again, see the boss and he will replenish. For those not fully conversant with First Aid, there's a book in the box. Extra copies are available. "Remember," said Robin, "The first man on the scene of an accident is always the most important man."

A word from Arnold Heine. Present temperatures and conditions make the ice cave near the base pretty unsafe. Best advice: if you don't want to become a sandwich under 50,000 tons of ice (useless), keep away. So says Arnold.

John Newman and Ian Stirling have lost some socks from the drying room. Check your supply and see if you have any spares that were not there a couple of weeks ago. Bill Lucy is getting frantic about his prized "spanner" (for opening beer cans), missing from the sledge room. It has a white handle and makes a big hole—in the can. The boss has mislaid his Sked book.

Lucy has designed a fascinating roster system for cleaning C Hut (large sleeping quarters) This will be put on the notice board later today and also in C Hut – when I get around to typing it up at my special rate of two cans of beer per copy. Check it to see when you are on duty. The hut mouse is responsible all week for general tidiness of the hut, entranceway and main base entrance. Each man is responsible for his own room.

[Saturday] November 30

Paul Gabites of New Zealand's External Affairs Department wrote a note of appreciation of his visit to Scott Base during a recent VIP trip at McMurdo. He says Scott Base remains the highlight of his visit south. "There are several reasons for this, some of which are not easy to define but which could probably best be listed under the Decent National Pride heading. The modest scale, the quiet but undoubted efficiency, the closeness of the challenge of the south and the acceptance of the environment and, thinking back, the absence of the spectacular and the presence and feeling of a great tradition are what distinguish in my mind your effort....the evening meal you gave us—three perfectly adequate pots on the stove ...."

Continuing news from Vanda is that the quintet team has moved into the living quarters that used to be the main hut at Cape Royds. Noel Wilson and his crew are now putting the joinery in the lab hut. The wind generator is now up.

[Thursday] December 5

The Flower Pot boys, Jim Cousins and Bruce Brookes, worked through the night on a seabed probe near Cape Evans. Before turning in this morning, they came on the emergency schedule to inquire the whereabouts of the icebreakers. Seems they do not want to wake up to find a ship in their flower garden. The icebreakers are not due in Winter Quarter's Bay until December 15. Yesterday they progressed 1½ miles through the sea ice in the vicinity of Cape Bird.

The Cape Bird team has sent a note back asking for a piece of 2 x 1 lumber about 10 feet long to enable them to carry out an experiment which allows penguins to look at their own reflections. So on the next flight up, probably tomorrow the cargo will include a Penguin Look Stick. Patents may be pending!

On Saturday Japan's Mr. Antarctica Dr T. Torii and Messrs N.Yamagata and Ichyou Mukou will arrive as guests but not members of NZARP. Transport will be arranged to get them to the Dry Valleys, and they will also travel to the South Pole, where they hope to meet up with the Japanese Antarctic Research Expedition traverse team.

Field assistant Noel Wilson checks field gear in his workshop.

[Monday] December 9

Work is proceeding at a torrid pace in the garage to get the Sno-Cat ready for two non-stop trips to the Bay of Sails with fuel for Vanda Station. The first party led by Bill Lucy is expected out tomorrow. About five helicopter trips will be saved by taking two loads of fuel to the Bay of Sails for uplifting by helicopter. Wayne Maguiness and Doug Spence will go with Bill while Hugh Clarke, Derek Cordes and Keith Mandeno will make the second trip.

Noel Wilson who left Scott Base with the original tractor train on October

23 will come out of Vanda today. He's been working on the construction of the new station.

Two albatross are on the US icebreaker Southwind from Campbell Island for shipment to Christchurch. Sending them south is the only way the birds can reach New Zealand before our supply ship HMNZS Endeavour calls in late January.

A well-known Scott Base tea drinker asks that the morning pot of tea be kept off the range after it is made. "Ah dorn't lak et stewed," he says.

[Wednesday] December 11

A weary trio arrived back at Scott Base around 5 am today after a non-stop round trip to the Bay of Sails with fuel for Vanda Station. Base Engineer Allan Guard was up at that time to carry out urgent repairs to one of the Sno-Cat tracks to have it ready in time for the second trip due to leave after lunch.

NZARP's three Japanese guests hope to meet the JARE pole traverse team on or about the 20th at the South Pole. Beforehand they expect to visit several places under USARP, including five days at Vanda Station, two days at the Don Juan ponds in the Wright Valley, two days at Cape Royds, two days at Lake Bonny as well as a trip to Cape Crozier on the eastern tip of Ross Island. They also hope to make use of the USARP fish hole hut to collect seawater samples. Journalist Ichyou Mukou will stay at Scott Base for about a month after the scientists leave.

The laboratory reports that a solar flare disturbance began at 4 pm yesterday. Only slight disturbance was noted in the ionosphere this morning.

Lab technician Keith Mandeno requests that the house mice refrain from turning on the lights in the P-R-O's darkroom as his plants cannot get any sleep with the lights on.

New Zealand's Minister of Agriculture Brian Talboys has sent two cases of oranges to the base. Robin has sent a telegram of thanks to Mr. Talboys, who is currently in the UK.

A party of seven US congressmen are due to visit Scott Base this afternoon.

[Thursday] December 12

Vanda Station is now on full power. Base electrician Chris Rickards reported by radio last night that the wind generator is now fully operational and providing 12 volts to the laboratory hut. Chris will get back to base on Saturday.

Even at Scott Base you cannot miss those exams. Alister Ayres sits his New Zealand engineering technician exam after Leader Foubister was approved as invigilator. Taffy passed the exam and post Antarctica gained qualifications as a land surveyor in both the UK and New Zealand. He finished his career as Chief Surveyor in the Cayman Islands.

Scott Base today established direct contact with the Japanese pole traverse party, now just seven days from the South Pole. The party left Showa Station on the Prince Olav Coast on September 26. They expect to reach 88 degrees south tomorrow when time will be spent on vehicle maintenance. On the 18th they will be 20 kilometers from the Pole that they hope to reach by 2 pm on the 19th. Dr. Torii spent 45 minutes speaking to the party.

While most of the base slumbered last night, our two grease monkeys were at it all night in the garage. The Sno-Cat had to be repaired in time for today's trip to the Bay of Sails. Wayne Maguiness and Allan Guard worked through to get the tracks to run properly over the pontoons. Wayne hit the sack at 3 am while Allan joined us for an early breakfast.

Birthday boy Bill Lucy (32 today), Derek Cordes and Bruce Brookes took off at noon hauling sledges loaded with fuel, spare parts for Vanda and the valley Ferguson and a knocked down refuge hut to be constructed at the head of the Wright Valley. Bill is not happy about the condition of the sea ice and just before he left, he asked the Post Office techies to maintain a continual radio watch from 3 pm. Derek and Bruce will remain at the Wilson Piedmont

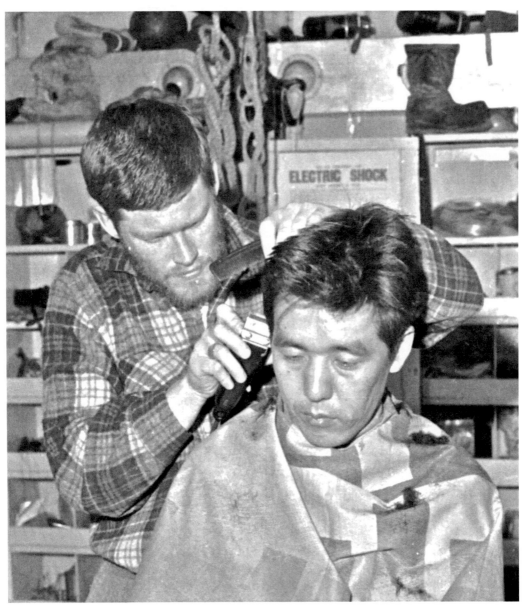

Volunteer barber Keith Mandeno tries his hand on Ichyou Mukou, a Japanese journalist who stayed at Scott Base waiting to link up with a Japanese South Pole expedition.

for a while to build a refuge hut and ferry the supplies up the Wright Valley to Vanda. Jim Cousins and Simon Cutfield will return with Bill in the Sno-Cat.

[Sunday] December 15

The Sno-Cat returned today after its fourth non-stop run with fuel to the Wilson Piedmont Glacier-Bay of Sails dump. The total fuel taken in the four trips is DFGA 45 drums, kerosene five drums and gasoline seven drums (each 45 gallons).

Cook Geoff Gill has received a pile of Christmas decorations from the Antarctic Division and will be putting them up soon. The New Zealand Antarctic Society is gifting a set of wineglasses for the wintering over party.

[Thursday] December 19

The first two permanent residents of Vanda Station, Simon Cutfield and Ron Craig leave Scott Base tomorrow. They will return to base next summer, maybe October 1969. Good luck fellas.

A story on Rugby practice last night was forwarded to the New Zealand Press Association today. The squads are: South: Lucy (captain), Foubister, Lennard, Spence, Newman, Guard, Clarke, Wilson. North: Mandeno (captain), Connell, Gill, Millar, Rickards, Maguiness, Blackbourn, Hancock. Managers: South Foubister, North Connell. Ballboys: South, Newman; North Gill. Refreshments, Foubister. Press coverage, Connell. Transmitting: Hool, Hancock. Referee: No appointment meantime. Clothing: South, mukluks and long johns; North, mukluks woollen trousers and bush shirts.

[Friday] January 3:

A story I sent out to the New Zealand media last month on the vast amount of philatelic mail handled through the Scott Base Post Office has reached the columns of the Johannesburg Star in South Africa. Today's philatelic mail included two letters from Johannesburg collectors. (I hear groans from the Post Office staff).

VUWAE 13 gained a day or maybe three. Dr Peter Webb and his university party were waiting at their pickup point on the Skelton neve today, a day earlier than their scheduled flight. They thought today was the 4th. The current program has them for pickup on the 6th.

McMurdo public works boss Commander Bob Booth called Robin today wanting to know why we were taking one of their Sno-Cats away with the D4 bulldozer. The incident was easily explained: The big Sno-Cat was gifted to Scott Base by USARP as parts for our Sno-Cat Able. The body will find a future as a ski lodge at the Kiwi skifield.

A brick for the happy whistler. Late house mouse Bill Lucy was catching some shuteye in his bunk this morning after coming off duty when: "Some crint came whistling and stomping round C Hut this morning. He left just in time...." So our deputy leader reminds all to think of others when going into the sleeping huts, day or night!

Report from Vanda Station today is that the Onyx River is now 18 inches deep and flowing at 10-15 cubic feet per second.

In preparation for the Governor General's visit next week, Leader Foubister has asked that from after breakfast on Monday till after breakfast on Friday we refrain from wearing our boots or mukluks in any of the huts. He suggests this will help keep the base clean, tidy and dust free.

[Monday] January 6

Rain fell in the Wright Valley today, and I sent a wee piece off to the New Zealand Press Association. Meteorologists Tony Bromley and Ron Craig were highly enthusiastic about this extremely rare happening in Antarctica.

They ran outside to enjoy the whole 15 minutes of it. Although there was only a trace, Bromley said by radio to Scott Base the rain was enough to put spots on the rocks. Similar rain was also experienced further up the valley.

Seismology lab man Keith Mandeno was ecstatic last night when he discovered an earthquake as he developed his seismograph records. He reckoned his earthquake to be about Force 7 or 8 with an epicenter somewhere in the vicinity of New Guinea.

Postmaster Brian Hool got on to his Wellington mates during last night's phone sched and found that the 'quake was indeed Force 7 and occurred in or near the Solomon Islands.

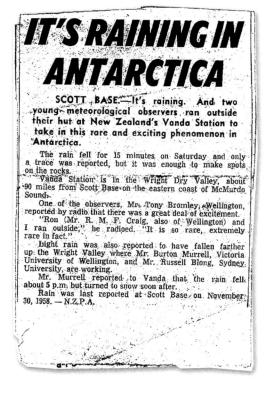

IT'S RAINING IN ANTARCTICA

SCOTT BASE.—It's raining. And two young meteorological observers ran outside their hut at New Zealand's Vanda Station to take in this rare and exciting phenomenon in Antarctica.

The rain fell for 15 minutes on Saturday and only a trace was reported, but it was enough to make spots on the rocks.

Vanda Station is in the Wright Dry Valley, about 90 miles from Scott Base on the eastern coast of McMurdo Sound.

One of the observers, Mr. Tony Bromley, Wellington, reported by radio that there was a great deal of excitement.

"Ron (Mr. R. M. F. Craig, also of Wellington) and I ran outside," he radioed. "It is so rare, extremely rare in fact."

Light rain was also reported to have fallen farther up the Wright Valley where Mr. Burton Murrell, Victoria University of Wellington, and Mr. Russell Blong, Sydney University, are working.

Mr. Murrell reported to Vanda that the rain fell about 5 p.m, but turned to snow soon after.

Rain was last reported at Scott Base on November 30, 1958. — N.Z.P.A.

Peter Webb and the VUWAE boys are still out on the Skelton Neve. They have been sitting, waiting, and tuning in on the radio since January 3. Maintenance problems with the Hercules C-130 that was to have picked them up today prevented us from seeing the gang tonight.

My newsletter quietly faded away on January 7, as life and work around Scott Base became too hectic with the Governor General's visit and the official opening of Vanda Station. With only about six weeks remaining for the summer programme, we moved into a frenetic pace to ensure all essential work was completed.

A Hercules C130 taxis at Williams Field, McMurdo, amid swirling snow.

8.
FIVE MINUTES AT
THE BOTTOM OF THE WORLD

The day of all days came very unexpectedly in November when I got the chance to fly to the South Pole courtesy of the US Navy for a story and picture on Arnold Heine, the NZARP glaciologist who'd been collecting ice samples each year for 12 years.

I had to be careful, almost nonchalant around the base as I was the first of our crew to get a flight to the US Amundsen-Scott Station at the Pole, each person's destination of hope. There was magic around the possibility of being able to stand at the bottom of the world.

The Hercules took off from Williams Field about 9 pm on what was expected to be a routine flight the 800 miles to the South Pole. I was the lone Kiwi on board and this was my first flight in a Hercules C130 aircraft. Warm, well-fed, comfortable even though my prime view from the folding military-green canvas seat was to stare at a giant black bladder fuel tank strapped down in the belly of this flying warehouse. We flew in the brilliant daylight of the Antarctic night at 26,000 feet above sea level (about 16,000 feet above the polar plateau). This was a rapid return shuttle flight, just cargo and four passengers. The fly boys told me we'd be in the air about three hours each way.

After about an hour of droning and staring at the whale of a fuel tank I went up to the flight deck to take a look around. It was a great viewing place but much of the landscape below us was packed in soft fluffy cloud. Here and there the cloud broke enough to see parts of the white mass of the featureless

Ross Ice Shelf. We'd passed over the point Captain Scott and his men found their final resting place on the return from their sad journey to the South Pole in 1912. The navigator pointed out our position as west of Little Jeanna, a US weather and radio station on the ice shelf roughly halfway between Scott Base and the Beardmore Glacier.

With the steady beat of the engines well behind the cockpit and seemingly nothing in front I had the sensation of floating. An ever-present Playboy pinup, in this case Miss January, was prominent above the flight commander's head. In this eerie calm, I smiled and reflected on the sign above the flight engineer's head: "Antarctic Flying. Hours and hours of boredom interrupted by moments of stark terror."

Around 10:20 pm the navigator pointed to the mouth of the mighty Beardmore Glacier, the gateway to the polar plateau for both the Shackleton (1908) and Scott (1911) expeditions. We were about halfway to the Pole as the Herc flies.

The Beardmore is wrapped in history. Shackleton named the glacier for a Scottish industrialist sponsor of his expedition. Located between the Queen Alexandra and Commonwealth mountains, the 100-mile glacier is one of the longest in the world. The early explorers saw it as an ideal place to move up from the ice shelf to the polar plateau. It became a place of struggle, privation and tragedy. The polar explorers made their way up with ponies and dogs and heavily laden sledges, faced blizzards, and struggled with inadequate food and lousy clothing. These tough eggs pulled their loads against time and weather. And Scott's pole party (1911), edging down the glacier on his return from dismal failure, was unable to find the outward route or the supply cairns along the way. Crevasses, frostbite and scurvy. How much did all that effort and sacrifice 57 years earlier contribute to my luxurious comfort in this flying machine clearing the same landscape in a matter of hours rather than the five months it took them on foot. These explorers, the leaders and, moreover the men, had broken trail for those who came later. Their passions, sheer courage and fight for survival lit the imaginations of later scientific endeavors ... and the adventurous heart of a dreamer in New Plymouth. The scenes of the 1948 movie Scott of the Antarctic had imprinted on my 10 year old mind.

We landed at the pole station dead on midnight, and in brilliant sunshine. Moving away from the plane I had a dazzling 360 degree view of nothing, just white expanse in every direction as far as the eye could see, a curved horizon. I was at the bottom of the world on a calm, warm day at -33°C. And, I was told, this gave the riggers the opportunity to work on the radio masts. My wide eyed wonder drained rapidly when I spotted Arnold waiting with all his gear stacked up ready for loading.

"Rats," I said to him. "Looks like I missed the photo op."

"We'll get about 15 minutes on the ground," Arnold said. "Go take a look at the Pole."

I'd been under the impression I'd get around an hour while the plane disgorged the contents of the fuel tank. My plan had included photographing Arnold collecting his ice samples and also a quick look around the American station itself. The station intrigued me as it had been constructed below the surface in giant trenches and then covered. Heavily-laden in down clothing and the cameras, I waddled-trotted some 50-100 yards with a US navy colleague to the shining Pole, a simple orange and black post in the ice with a shiny silver ball on top. Next to it was a standard highway-style signpost indicating mileage to major cities and places north in any direction. The Siple Pole was named for the American cold weather scientist Dr Paul Siple who was chief scientific officer for the first winter-over party at the Pole station in 1957. He is known for coming up with the term wind chill, the combined effect of temperature and wind on the skin.

I had taken four frames when I heard the Herc engines starting to scream. "Oh, no," I hollered. "I'm supposed to be on that." I dumped the cameras into the bag and sprinted to the plane. Out of breath and with my head spinning I mumbled into the frosty air, "All the doors are shut. I've missed it."

Then the engines slowed. Two men at the taxiway fuel line grabbed an arm each and hustled me over to the Herc. The door opened, the ladder dropped and the loadmaster beckoned to hurry. Did I ever. I reached the ladder and collapsed. With my helpers pushing me from behind and the loadmaster hauling me up, I landed face flat on the deck as the door slammed shut. The Herc was now on its way. I was helped to a seat, buckled in and asked if I

Glaciologist Arnold Heine (left) and surveyor Charles Hughes.

needed oxygen. "I don't think so," I muttered, noting the glassy stares of my new traveling companions. "Why?"

I'd forgotten that the Pole station is around 10,000 feet above sea level and the air is pretty thin. I had run at full tilt in heavy clothing plus about 15 pounds of camera gear. I was simply blown out.

I'd almost been marooned at the Pole. But that would not have been a bad thing, I thought. The station of huts buried under the ice would have been a nice place to stop over, and the folk there would have really looked after me. But the big question always was when the next flight out would be. As it was our plane had to take off quick as the forecast for the return flight was not good and conditions at sea level at Williams Field were deteriorating. My time at the Pole had been about 10 minutes. I didn't even get time to read all the destinations on the signpost!

Once I'd recovered, I had a great time making notes as Arnold told me his story of 12 summers on the ice gathering snow and ice samples for later analysis at the lab back home in Wellington. There would be another opportunity for a picture of him in the field closer to Scott Base.

November 20, 1968

Darlingest

Sometimes I have heard you talking to me. And the girls as well. Your voices have sounded as though you are all right in the room with me.

I have sent you some more photos. Labels are on the back. Be a good idea if you would wash them again. I am not too confident about the system here. I use Permawash. From fix to water for two minutes, then Permawash for two minutes and back to water. Thing is the water is supposed to be running but that is out. So I soak them well for about 10 minutes in as much water as I can afford then a final Permawash for about 15 minutes then back to water for up to half an hour.

I have only had one opportunity to go skiing. And I qualified for membership in the Scott Base Ski Club (a successful run down the hill!). Today I was admitted as a member of the Antarctic Press Club, a US outfit.

A word about meals:

You asked me about meals the other night. Meats range from chicken, lamb, ham, pork chops, lamb chops, steaks to liver. Veggies include brussel sprouts, broccoli, beans, peas, carrots, plastic spuds. We have salads about twice a week and deserts are usually fruit and instant puddings or ice cream or custard. For breakfast, I have porridge and then maybe eggs, bacon, spaghetti, baked beans, whitebait fritters, corn fritters, that sort of thing, topped off with a glass of orange juice and a vitamin pill. Coffee, cocoa or Milo (malted hot chocolate) follow. We always have a cooked meal for lunch and another at night. Sometimes there is wine with meals. After supper we either go back to work, do some outside work when necessary or just lounge around drink beer and yarn, play cards, darts, draughts, go walkabout, or enjoy a movie (Thursday and Sunday).

Keep warm

G

November 1968

Hello Sweetheart:

The girls are extremely good and help a lot. Rachel is growing up nicely. Hilary got a thrill last Sunday as she found out she got third place in a fire prevention week colouring competition. She burst in to our bed at 6:30 am saying: "Mummy, mummy, my name is here. Her prize is a $2 book voucher. Bridget came out with "good girl" tonight. Hilary very proudly read your letter to them. I didn't read for her. She made a good job too. Rachel likes your frozen rocks.

Hilary had a lovely time at the Fire Station. She sat up front with the driver and looked all royalty. The fire engine was loaded with children. We went straight down to the bookshop and Hilary ended up with four books and I bought two for Rachel as well. We caught the bus home loaded with parcels. Our two girls are a real joy to take to town. Gin looked after Bridget and I looked after theirs on Saturday.

Bye Sweetheart

Tuppy

December 7, 1968

Hello Hon:

Well, here ah is again, full of wonderful thoughts of my darling and her three charges. I am night watchman again and have now finished my job apart from getting the tables set for breaky and waking the mob. After breakfast, I'll be off to bed till lunchtime. If I can get the letter completed now I'll be able to get it in the mail that is supposed to be leaving tonight. Flights are getting more unpredictable these days. We never know when we might receive or send mail. I was right about the phone call the other night, they were charging. $1.80 would you believe? That will run away with my funds. I have sent another little surprise for the girls...a Kennedy half-dollar an American pilot gave me for them.

I have just been on my rounds and decided I am hungry. I have a piece of toast on top of the range, lotsa butter and Chesdale (processed) cheese washed down with perked coffee. Can't stand that instant stuff anymore. I've got Patrick O'Hagan (an Irish tenor) on the record player. We've played all the others. Our new records don't come in until the ship arrives in early January.

Allan Guard, our Base Engineer, gave us an illustrated talk last night on his 1965-66 year at the Campbell Island weather station. It is a beautiful place. Bird life there is fantastic. He had some damn good slides, too.

When I get Hilary's letters, she is right here beside me. Her little tongue poking out, freckles on her nose. And Rachel, it was lovely when she spoke on the phone. They are certainly not forgetting their father. Sometimes I think I was wrong to desert you all for this time. Look at Bridget, what am I missing... what a pity Hilary got so sunburned so early in the summer. You must be having some wonderful fine weather up there. Some days, and today is one, we get up here and I think of summer

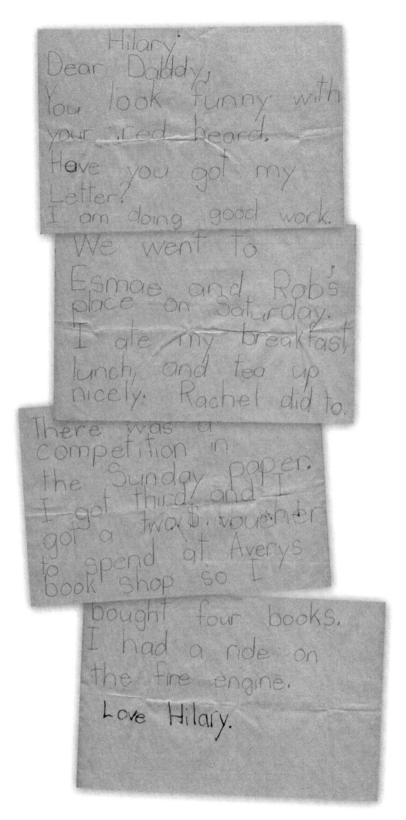

A letter from Hilary.

mornings and that feeling of wanting to get down to the beach. I suggested that to John today, that we get our rods and toddle off down. The others think we are crazy when we talk about acting out these fantasies.

No plane yesterday and it looks like no plane today because of radio blackouts. I got up at 11 am today as we are allowed to do what we like on Sundays. After lunch, I've got six or seven films to process so it will be good to get them out of the way.

Last night we dressed up and went over the hill to McMurdo to a chief petty officers' bunfight. Quite good. They have a barbecue there! Actually it is outside and they have a small hatchway into their mess. As well as the beautiful steak, we had chips (fries), baked beans, chili sauce, shrimps, pickles (gherkins, cauliflower, grapes and onions), potato crisps and a cheese dip bowl.

Someone brought out a guitar but no-one could play. Our postmaster Brian Hool got hold of it and led a couple of numbers and so it passed to me. Well, who said I could play? I had those same chords of old and a good strum so away we went, everybody singing at the tops of their voices. I must have been ok as I heard Robin declare "We've got talent" to which our hosts said: "I thought you said no-one could play?" Then they all wanted me to play their favorite songs while they sang solos! Whew! Must have been the power of beer and whisky.

I posted three big gifts to the girls yesterday. They are Christmas presents, but it is over to you when you let them open them. I thought I had better send them off while the going is good. They are penguins. Next week when I get some money I will send them a bracelet each. (Items purchased from the US navy store at McMurdo)

I will be phoning you on Christmas Day. At the moment it looks like anytime between 9 am and Noon. This has not been confirmed. I want to be first on at 9 am. So far, my name is at the top of the list. Let me know your plans. You could send me a comb for Christmas then I can comb my beard. It is getting a bit thick for my brush.

Robin reckons things will quieten down a bit in the next month. Perhaps I'll get a chance to take and process some pictures for us.

We have a great record down here. In fact it has been played so much that someone who did not like it went and hid it. Would you like to buy it for yourself? I would like you to then I'll be able to remember my sojourn down here and all the people. And maybe when you play it you will think of me. The record is "Look of Love" by Claudine Longet.

Your best bud,

G

December 15

Hello My Darling Sweetheart:

I rashly promised Hilary she could have a new dress for the prize awards ceremony at the fire station. So what with mine underway for the Journalists' Christmas function, and one of Ally's to fix, I have been sewing flat out and they are now all finished. I managed to squeeze a new one in for Rachel as well. They look lovely in them.

Hilary says you are a bad Daddy for wearing your slippers in the snow, Rachel has packed her suitcase and heading for a week with her cousins by the beach. Rascal Bridget is into everything. She now has a repertoire of: Look, see, please, thank you, good girl. She's coming along. She pointed to a picture of you and Chippy and declared: "Daddy, see". She is very pleased with herself.

Rachel came running in the other day with two letters from

Scott Base. I thought, ooh super and eagerly started to open them. What's this? Rachel Connell...so I tried the next one, Hilary Connell...and none for me. They were thrilled of course that they got letters and mother none.

We have an amusing time in the shops, the girls, mainly beaming Rachel, say proudly: "My Daddy's at the South Pole". Invariably the shop assistant looks at me unbelieving only to see me nod and smile. Much abashed, they say "Oh!

L'amour,

Tuppy

December 17, 1968

Dear Darling:

I had a bank statement on Saturday to say I had overdrawn, much to my amazement. It was through a misunderstanding. When I phoned to see if the money was in, I was told yes, $130. But when I get the statement it says $100.30 so that is how I was out. Your payments to date have been $170, $117 and now $100. Bit odd don't you think? But don't let it worry you, Sweetheart, as I'm not.

This weekend has been busy as I had my Dad over building me a gate for the driveway. Next weekend he'll build a small one for the front path. My design looks really superb, rough sawn timber painted white. He came over the next day and Hilary asked him to stay for supper because then the spare seat ("Daddy's," piped up Rachel) would be filled.

On Friday we bought some plants for the veggie garden and in the afternoon we all went down to plant. Or, so I thought. We all ended up doing different things. Rachel squirted water everywhere but the right place. The hose split and I then had to fix it while it was still going, so she ended up drenched. Hilary managed to plant a row of carrots. Bridget looked like the original mud pie baby. There was mud everywhere. The garden got a really good soaking. Bridget just sat and played in it as though it was sand at the beach.

On Wednesday, Mother found my ring! It was in one of the scrap parcels that had come apart on the back seat of her car. Wasn't I lucky. So now I take my rings off when I do washing, gardening, or anything to do with water. Phew! I was truly pleased the ring was found. I'd been feeling quite dreadful and even kept hiding my hand so I wouldn't be reminded of what I'd done.

Howard, Gin and all the kids and dog took off to the beach at Ahu Ahu Road last weekend. I borrowed a tent from your Dad. We went early Saturday morning and stayed till late Sunday afternoon. It was really wonderful, very quiet and lovely. Hilary is terrific in the sea, out in the waves all the time with me trying to teach her how to surf in on them. She is quite at home in the sea. Rachel is not as keen and on Sunday she slept until 9:30a.m. after having had her breakfast at 6:30 am (we were all up at 5:30 am!). We built a campfire for our stove, then we tried to catch some fish. Howard broke his reel and I lost a set of hooks and a sinker. No fish. We all felt as though we had been out there for ages. Crazy, isn't it, how time means nothing when you are at the beach? I gave Bridget a short haircut and she looks beaut. She takes to camping life very quickly and is no trouble at all. Mind you, when it came to bedtime, I did have to hold her down in one place long enough to get her to sleep. Then she slept like a log.

Back home again, and Rob and Esmae and the kids came for a visit and just as we sat down to lunch my folks got back from their holiday. They went right up to the top of the North Island and Ninety Mile Beach. Mother reckons that the north is the place for us. We could open a little shop, even a hotel or motel. She said it would be terrific for us and she thought a lot of us, driving around. There are lots of old things being discovered and restored which is right up your alley. Perhaps you could go native and take up fishing, carving and selling books.

Rob mowed the lawns for me this afternoon and then he took me to the movies leaving Esmae with their five and our three! We met some of the guys from the Herald and they wanted to know if you were going back there. They reckon the place is pathetic. I simply said I did not know what we were going to do. I just don't know what we are going to do. We'll just have to wait till you get home before we can really decide. One part of me wants to go abroad or move, and another part wants to stay. I don't know which one will win!

Bonne nuit mon petit chou,

Tuppy

As a well-bundled polar photographer, I try for some good scenics on the sea ice of McMurdo Sound.

9.
PLAYGROUND OF THE WIND

I felt a bit like a wad of scrap paper, tossed from my typewriter, slowly un-crinkling close to the waste basket. I blinked, blinked and blinked again. My eyes were crusty from sleep and really didn't want to open. Besides, where the blazes was I? Not in my bunk that's for sure. My back ached but not enough to make me move. Snug and warm and half asleep, I figured I should stay that way. My brain creaked to action. My eyeballs slid slowly, side to side. I found the face hole of my sleeping bag, heavily lined with the frost of my breath, and peeped out. Beyond the frost was daylight and it slowly dawned on me I was enjoying a new campout in Antarctica.

I dared not stir in spite of the soreness from stiff muscles seemingly attached to the lumpy, frozen ground. I lay there on my side listening to the wind and the sand raining down on the orange canvas shell of my bed. I peeped out the face hole again to spend a moment or two marvelling at the intricate patterns of gravel and brown sand embedded into my ice window.

The events of the past 24 hours crystallized as I eased very carefully on to my back hoping the frost around the mummy bag face hole would not result in the chilling discomfort of rolling down my neck. I looked straight up at white. That had to be sky, I thought, because I am lying on my back.

What else is there? Ever so carefully I brought my left arm up to where I could see my watch, 2 o'clock. Is that morning or afternoon? Day of the week? I played with this for bit and reached agreement with my sleepy self that it must be Tuesday morning November 26. What did it matter anyway? I was marooned some 50-60 miles northwest of Scott Base at the Strand

Moraines, a dirty tongue of gravel and ice at the foot of the Bowers Piedmont Glacier. I thought about how we realized that somewhere in these gravelly hummocks, we'd build our "bunkroom" to get out of the wind. We'd spent a good bit of the afternoon roaming over and down a few hummocks to select a relatively sheltered spot to set up camp in the desolate glacial debris left from a previous ice age some 20,000 or maybe 50,000 years ago. We'd been warm and cheerful. It was surely one of those never-to-be-forgotten you-should-have-been-there situations we'd be able bore the rellies and friends with for years to come.

As I wriggled around to get a better view out of the sleeping bag face hole, ice flaked off, fell around my face and neck and melted into icy dribbles. Not much to see except a dirty ice wall. Grabbing hold of the inside of the bag near my neck and by tucking my legs up, I managed to roll the whole bag over to scan in the opposite direction. There were my three camp buddies bedded down in the lee of our crude, wind-breaking, ice-walled bivouac. In the wider view, squawking skua gulls stood sentinel on the surrounding gravel hummocks, eyeing our orange intrusion into this frozen moonscape. Every now and then, one of them swooped low and hovered over us. There was no way this shore-based predator of the Antarctic would find an appealing meal in orange canvas. But then, these critters are mean; they might try anything! I'd watched them attack chicks and eggs in the penguin colony. Baby seals are not immune to their cunning voracious attacks. As chilling wind whipped up the sand around our shelter and rained down on our sleeping bags, I snuggled back inside.

What had started out as a simple five-hour Canterbury University (Christchurch) seal-tagging trip had become a night in the open. Conditions locked us into the Strand about noon, and the whiteout prompted the abandonment of our scheduled helicopter pickup. Once again, this wonderful yet strange South Polar Region was in control. My short trip for pictures and a feature story for a Christchurch newspaper had now evolved from a factual account into a personal experience feature story and pictures for the press association syndicate. I laughed to myself at the irony, rolled back on to my left side and stared at the dirty ice wall. I was surely in the best place under the circumstances. I worried I might have to extricate myself from the warmth of my bag to go out for a pee.

On the far side of the bivouac was Alessio Ollier, one of four Italians working with this year's New Zealand Antarctic Research Programme. This would be a new experience for him, a veteran professional alpine guide from the Italian Monte Bianco (Mont Blanc). Next to him was David Greenwood, a museum technician with the Canterbury University zoology department. Like myself, he was an Antarctic neophyte and an aspiring OAE (Old Antarctic Explorer).

Bivouac in the Strand Moraines. From left, Alessio Ollier, Dave Greenwood, Ian Stirling and my empty bag.

Snoring happily alongside me was Canadian Ian Stirling, a zoologist at Canterbury University. This was a been-there, done-that yawn for the four summers man. The same thing happened to him the year before and in much the same location.

My involvement in this field trip was to record in word and pictures an article on Ian's work on the ecology and population dynamics of the Weddell Seal. David, as his understudy, was expected to continue the work next year in the development of a benchmark should exploitation of the wonders of this part of the world ever become de rigueur. This summer's studies would close Ian's work, leaving the way open for the program to be set up on a long term basis to keep track of the survival and dispersal of marked, known age groups, the result of tagging pups each year.

It was cold work with the icy wind roaring northward off the polar continent, freezing bare fingers in seconds. It took five hours for the Ian and Dave tag team to recruit 21 new seals, document two females tagged the previous summer and check out three dead pups and

Deplaned, with our survival gear and ready for work.
From left Alessio, Dave and Ian.

a dead adult male. We spotted one candidate out on the pack ice and enjoyed a 45-minute walk each way to enrol him into the programme.

The area attracts its own Weddell seal community. The same happy chappies come back to breed each October, year after year. We saw about 30 compared with 45 at the same time a year before and 45 the summer of 1966-67. Heavier fast ice probably affected the resident population. Thicker ice from cooler temperatures coupled with the lack of high winter winds to break it up tended to spread the seals further north along the Victoria Land coast.

This beautiful 1200 pound silver-grey animal lives a catch-22 life. It has to get into the sea to feed and it has to get its nose out of the water to breathe. Often these Weddells maintain their own air holes by gnawing the ice to keep them from freezing over. Their task is made much easier if they can find gaps caused by glacial movement and tidal action. Probably easier on the teeth, too.

A seal pokes his head out of his breathing hole in the sea ice to see what is going on.

To keep his subject calm during the tagging operation on the tail flippers, Ian and David pulled a grey canvas New Zealand Post mailbag over the head. The poor seal then just lay there unmoving until the measuring and tagging was over. Once the bag was removed, the seal would take a look at us, roll over, and maybe give a short grunt-bark before getting on with what he was doing, snoring off his last meal.

It is easy to form a liking for this massive 10-foot long seemingly helpless animal. They have a beauty all their own as they lie dozing not far from their dive hole, the sun glistening on their sleek silver hide. They spend most of their time under the ice where the chilly waters of the Ross Sea provide all the food they need, cod, silverfish, crustaceans, octopus and other marine life. They can dive to something like 2000 feet and stay under water for about 45 minutes. Their biggest worry over their 30-year life comes from orcas that like to include tasty seal in their diet.

Dave and Ian bag a seal to keep him still and quiet for measuring and tagging.

I was amazed at how I could just walk up to and ever so gently touch one with my boot before stroking its coarse, silky-looking, fish-smelling fur. They might roll over or lift their heads to give a baleful look at any two-legged, orange-trousered intruder. Sometimes I got a bit of a bark from one or a hissy yawn as they lay around the pack ice alone.

A Weddell youngster.

The world's most southerly occurring mammals, Weddell seals are plentiful in the Ross Dependency because of the abundance of fast sea ice. While they live right round the massive Antarctic Continent, the greatest concentration is between Scott Base on the southern tip of Ross Island and Cape Royds, some 25 miles up the western coast, the site of Shackleton's 1908-09 headquarters.

Ian reckoned the Weddell was the best species for a study such as his as they are not afraid of man, are easy to catch and tag, and occur in sufficient numbers to gather a lot of quantitative data. The geographic location is also ideal with close logistical support from Scott Base and McMurdo Station. The study over the past four years also included pods at Cape Hallett and northwards along the Victoria Land coast of McMurdo Sound.

Before he came to New Zealand, the 27-year-old Canadian had worked in the Arctic on bird and mammal studies for the National Museum of Canada.

This would be his final season in Antarctica.

With the tagging work completed, we'd concentrated on finding some way to get out of the incessant wind and build a bivouac for additional shelter. I found we'd have to be pretty creative as I figured just getting in the lee of a hill would be enough. This was not the case. The wind just kinda roller coastered up the giant gravel heaps and gathered greater momentum for the downward spurt.

Ian Stirling tags a Weddell seal.

The Strand Moraines is certainly a weird place located on the Bowers Piedmont Glacier, south of Butter Point and south of Cape Roberts and the Wilson Piedmont Glacier where I'd been a few weeks before. The Strand edges the sea, a barren waste of dust, and gravel, and ice fused into wind-shaped hills and hummocks. It is nothing more than a grey-brown desert frozen in time.

Our Canadian OAE's similar experience in the Moraines paid off and we set to hunting around for anything that might be used to fashion a shelter for some respite from the biting, howling wind. We had to stay close to the shoreline in case a chopper came in unexpectedly as it would be difficult to find us near a suitable landing spot amongst the hummocks. We built the three-sided bivouac from chunks of frozen ice and gravel we manhandled from far and wide. With our ice axes we chipped, chopped and then rolled or lugged chunks of gravel-filled ice from wherever we could spring it free from the tentacles of the permanent ice. We called it quits when the walls reached about three feet in height. We'd left our orange survival bags about 150 yards away out on the sea ice just in case the weather lifted sufficiently for a helicopter to come our way.

We spent the rest of the afternoon exploring in our gravel pit. In one hollow, we found a beautiful ice-covered pond. The surface was an intense pastel blue. We lay on our stomachs and gazed at the myriad of flower-like patterns embedded beneath the surface.

Ahhh, summertime on the ice of McMurdo, a Weddell mother and her pup.

To keep warm and occupied in the bitter cold afternoon wind, we played ice hockey using ice axes for sticks and a small stone for a puck. Tiring of that, our Canadian set us up with a curling rink. I was enthralled with the tales he told about growing up in Kimberley, British Columbia. Perhaps, just perhaps, Lois and I and the three girls would get there one day, but how to put the wherewithal together on a journalist's pay escaped me. Only dreams are free.

Alessio had joined in all the games throughout the afternoon. The language barrier made it difficult for him to understand us, but as an elite guide and alpinist he must have guessed. About 5 pm we returned to the bivouac after collecting the gear from the sea ice. This would be home for at least one night. As we broke out the survival kits and prepared our beds, Alessio grinned, finally grasping the situation and realizing that the helicopter would not be coming to get us. He now saw the bivouac had a purpose and was not part of our games afternoon.

Supper in the cold grey of Antarctic nothingness consisted of trail biscuits, delicately laced with lashings of peanut butter, jam and cheese. Dessert was a handful of raisins and half a bar of chocolate each. Very thankfully, we'd saved one of our lunchtime thermos flasks of coffee.

A radio aerial was rigged between two ice axes on the "patio" of our shelter. We tried to call Scott Base at 6 pm when we knew a weather report would be made from Vanda Station. Neither Vanda nor ourselves could raise Scott Base so we cheerfully talked with the Vanda crew for a few minutes. We tried the radio again at 7:15 pm on the emergency schedule but again no success. In the daylight of midnight we did the same. It was another Vanda weather reporting time, but although we could hear the base operator clearly, neither Vanda nor us Morainers could be understood. The night had been long and the never-ending moaning wind showed its eagerness to get to New Zealand

A heavyweight Weddell seal says hello.

by whipping across our little shelter, peppering our orange sleeping bags with sand. The heavy polar down-filled bags were enveloped inside a canvas outer bag. I was glad I had picked one end of our bivouac, for when I rolled on to my left side I could imagine I was inside a building trying to look through a very dirty window.

Radio contact was established with Scott Base around 8:15 am. We called again at 9 am to learn that a chopper was on its way. We figured we had almost an hour's worth of bag time up our sleeves. Just minutes later, though, and without warning, it was chaos. The darned helicopter popped up over the hill behind us glaring like some giant orange-faced locust. Our rescuers circled round and landed, scattering grit and dust over our little camp. We scrambled to; dressed and stuffed our bedding and gear into the survival bags and bounded over to the whirling Sikorsky. The loadmaster cheered and pulled us up and in.

Our 25-hour adventure campout was over. The engines revved, the blades whap-whapped as the machine rose in a swirl of sand, twisted round over our lunar gravel pit and set a course across the sea ice to Scott Base.

One of the joys of having three daughters and living close to the sea is picking around amongst the rocks at low tide. For Lois and I, this was a very inexpensive way to spend an afternoon, summer or winter, sunny or stormy. The girls loved to fossick in the tide pools and always, regardless of which beach we went to the rear floor of the car would be littered with rocks and shells. And if we headed to the mountain walks, I could just about guarantee the girls would find a nice looking rock or stick that just had to come home with us. The attractions in a rock for them could be just the colour, how smooth it was, shape, or whether it just had interesting flecks in it. So it was little wonder that wherever I went in my travels around the ice, I picked up bits of rock and took them back to office to put on a shelf beside my desk. After several weeks of this the little row of rocks was starting to take on a life of its own as the rocks were different from anything I had seen around the mountain streams and beaches. Our carpenter John Newman noticed my collection and confided he'd been doing the same thing. Our jackets always contained a rock or two when we came back in from the field. My collection was for no other reason than they were Antarctic rocks. I had the notion that I might send them to the girls so they could take them to school for some sort of show and tell.

Over a beer one night, John had the idea of building a box to glue the samples in for possible school talks when we got back home. We agreed on a joint effort and with his skill as a carpenter, we set about crafting a box each 3.5 inches deep, 18 inches wide and 34 inches long. He made it with a groove in the top for a sliding lid.

My Antarctic rock collection safe in a 34 inch by 18 inch display case.

The box made for a great diversion in our work-sleep world and we'd spend time together in his carpenter's shop building and finishing the boxes ready to accept our samples as we gathered them on our various journeys.

On Boxing Day, I wrote to Lois telling her how much of a pack rat I had become with various souvenirs and John's effort to make me a display case. "What with my rocks," I wrote, "I now have two penguin flippers, a skua gull bone, seal tooth as well as the stuff our Japanese visitors have given me. Our foraging habits have now encouraged others into the collecting game."

By January we were well on our way with our collections. We had to get a move on though to complete the carpentry. Robin gave me a map for the box, which I pasted onto the inside of the lid and marked where I'd found the various rocks as best as I could from memory. Each sample was numbered with a reference key pasted on to the lid.

"John and I sneak off after lunch down to his workshop to work on the boxes," I wrote to Lois on January 18. "Mine is going to look quite good. I think I will end up with 50 or so samples by the time I'm finished."

A month later I had all my samples glued into the box and put numbers beside each rock. Like our girls, I had based my collecting around colour and texture and just plain different looking stuff.

I vowed to keep the box beside me on the way home and told Lois so. I was quite proud of it, as apart from photographs, it would be the most tangible thing I'd have to show for the greatest adventure of my life. I knew deep down that as sure as my summer days were giving way to night on the polar plateau my little rock box was something I'd always be able to touch. Besides, I'd convinced myself that people would want to check it out and that the girls would refer to it at their schools.

Together John and I tackled the Victoria University geologists when they got in from the field for some real help to identify the rocks. Their leader, Dr. Peter Webb, got very excited over the collections and from my prompting, found the time to give the whole base a geology lecture one Saturday afternoon using the boxes as part of his talk. He also gave us a couple of interesting samples to add to our collections.

I had a great time that afternoon as the geologists from Victoria pored over our samples deciding how to describe the rock. Of course, being total amateurs, we had not noted where we had found the piece other than the general vicinity, like Cape Evans or the Wright Valley or Vanda.

The rocks in the box:

1. Basalt scoria – general area of Scott Base and Crater Hill.
2. White marble – Marble Point.

3. Beacon Sandstone – Boomerang Range. Note iron minerals; laid 300 million years, contain fish fossils from late Devonian or Carboniferous period.
4. Pure Beacon sandstone. This one ortho-quartzite – mainly quartz grains.
5. Amphibolite, a basement rock from the Wright Valley. White is feldspar.
6. Quartz veins in wall rock, basement.
7. Porphyrym (dyke) thin sheets incline at angle; white is feldspar crystal.
8. Ortho-quartzite, from a Dry Valley.
9. Jointed fragment dolerite (fine grain) common in Dry Valleys.
10. Quartz pebble – derived pebble, long time to round.
11. Typical granite from Dry Valleys. Pink yellow crystals, feldspar; transparent, quartz; also small grains of mica; tiny shiny surfaces, hornblende.
12. Basement schist of Asgaard group. Found at The Strand Moraines.
13. Fine grain basalt from Dry Valleys.
14. Ortho-quartzite.
15. Porphyry-dyke—basement rock from Dry Valleys.
16. Basic nodule from basalts. Pale Green, olivine; dark green, pyroxene.
17. Dolerite.
18. Granite porphyry showing feldspar crystal.
19. Dolerite sill—coarse grained.
20. Quartz.
21 & 22. Scoria from Crater Hill, red and black color depends on state of oxidation.
23. Fine grain granite.
24. Gneissic granite, basement rock.
25. Granitic basement.
26. Granite type beacon sandstone, contains 'cooked' quartz and feldspar.
27. Basic scoria from Castle Rock, collected ash en route.
28 & 29. Red and black scoria, orange has weathered.
30. Olivine nodule in black scoria.
31. Basement rock not known, well weathered.
32. Gneissic granite.
33. Heavily weathered basalt rock from Castle Rock.
34. Prismatic vein calcite (unusually good specimen).
35. Basement contains feldspars.
36. Granite, good specimen contains more quartz than normally seen.
37. Kenyite lava from Cape Evans. This Kenyite only occurs on Ross Island.

38. Metamorphic rock, still sedimentary layers; many badly buckled.
39. Gneissic granite "chewed up".
40. Glassy scoria showing fibers from top of Mt Erebus.
41. Petrified wood found in Mt Fleming area at the head of the Wright Valley.

A chunk of petrified wood given to me by Italians Marcello Manzoni (geologist) and Ignazio Piussi following their explorations in the Mt Fleming area at the head of the Wright Valley.

Bill Lucy, deputy summer leader and Vanda Station leader, on the flight deck of the Herc as we fly around the Boomerang Range.

10.
A FLYING ASSIGNMENT

For the second time in a month, I was on a Herc, this time heading out to the US Byrd Station, about 1000 kilometers east of Scott Base. My flight to Byrd was a roundabout route for a reconnaissance flight southwest of Scott Base over the Boomerang Range area where the VUWAE 13 (Victoria University of Wellington) geologists would spend several weeks looking at rocks.

The flight gave me the opportunity to get good info from the team leaders for a feature story I was writing. Heading the team was the university's Dr. Peter Webb, a veteran of Antarctic exploration. The main reason I was on board, though, was to get aerial photographs of the polar landscape in the vicinity of where the group wanted to continue their Beacon sandstone geology studies. My pictures were to provide a good idea of how a giant Herc could plop down out of the sky, ski to a successful landing and drop off the scientists and their gear.

A photo record of possible landing surfaces was surely good planning, and I felt I had no alternative but to turn in a value-added portfolio, though that seemed an impossible task with just a basic Rolleiflex camera and my Canon FT single lens reflex.

After boarding the plane I did not have a clue as to where I could shoot from. There were only a couple of eentsy, weentsy windows somewhere near the wings in the cavernous interior.

I'd realized after several weeks into my summer adventure, the just wing-it-and-do-what-you-can philosophy was a generally accepted rule of exploration in the extraordinarily beautiful, rugged, dangerous, awe-inspiring Antarctica.

The joyride to Byrd would take about three hours. I knew I would not de-plane there, which was a pity as I wanted to see Byrd, an interesting manmade research centre below the featureless plateau. The base had been officially brought into service some six years earlier and represented a unique approach to living and working in such an inhospitable environment. Earthmoving equipment had carved huge trenches in the ice cap and these were roofed with steel arches. Oil-heated buildings were constructed in the trenches with air space all round for insulation. Not far away from the station, scientists recorded the greatest ice thickness ever measured at something more than 14,000 feet (2.75 miles), depressing the land itself to well below sea level. An ice core-drilling machine brought up samples of ice at Byrd as old as 50,000 years. Scientists can gauge the age of the ice by the number of layers, each layer representing a year. A bit like a tree really.

What made the trip to Byrd even more interesting was that a couple of days earlier we'd had a visit at Scott Base from Tony Gow who told us all about the core drilling programme. Tony was a New Zealander, a petrologist and former lecturer at Victoria University, now working with the US Army Cold Regions Research Group. By the time I met him, he'd been coming to Antarctica for 11 summers and was back at Byrd on the drilling programme into the ice cap. It was tough going for the drillers because of the extreme cold at the bottom of the hole. Drilling fluids kept freezing and jamming the bit so they were on the forefront of developing the technology to keep going.

As we droned across the barren Ross Ice Shelf, I thought back to 1961 to one of my biggest and earliest flying adventures, the day Lois and I boarded a swept-up DC3 to go on our honeymoon. The NAC (National Airways Cor-poration and forerunner to Air New Zealand) operated out of a green cor-rugated iron lean-to attached to a World War II hangar at the original New Plymouth Airport. That Sunday morning after our wedding, both of us were dressed for our royal occasion Lois in her red going away outfit with beige hat and me in a dark grey suit. Surrounded by family and friends, we were able to go out to the plane and pose with our baggage in the sunshine under the nose of our shiny silver transport into the future. On board and all belted up, Lois grabbed my hand as the piston-engine plane rolled out to the grass runway, revved its engines to howling point, shook and rattled almost to bits before surging forward, lumbering down the green, green sward, and lifting its tail to catch a wave into the sky. The plane circled round the airport,

Flying 1961 style, a DC3 will take Lois and I on our honeymoon.

tipped its wing to the mountain and headed south to Nelson at the top of New Zealand's South Island. Lois was flying for the very first time, on our way to a couple of weeks at a beach cabin loaned to us by one of Lois' uncles. It was early Autumn and the magic of Fall colors and sunshine seemed to be preparing us for our life of living together.

In the Herc we all soon took off our heavy down-filled clothing, which some rolled up as a bed on the floor for a good nap. About an hour out, the weather cleared and, as on my previous flight, I went up to the flight deck for a peek at the outside world. I got to sit in the pilot's seat (its o.k. fellas, the co-pilot was at the controls). I felt like I was quietly floating, weightless, over the all-white landscape of the Ross Ice Shelf and into amazing clear blue space. I put on a headset and adjusted the microphone. To talk with others on the flight deck, I flicked a small switch on the side of the joystick. I recalled my days as a teenager lying in the tall grass in the fields back home to watch Cessna aircraft and Tiger Moth biplane topdressers spread fertilizer on the steep hill country sheep pastures.

"I wish we had something to do," the navigator remarked. " This is so boring." After the commander told me to take a look at the contrails, our laconic navigator chipped in, "What's up, we lost an engine or somethin'?"

We were reminded a couple of times that we'd touch down, unload and be gone again within 15 minutes. No one was to leave the aircraft. The plane's crew treated us royally and kept plying us with coffee. The ever-present coffee pot was something new for the tea drinkers of New Zealand. And the bottomless cup was very easy to get used to.

We were very fortunate to get this flight as all US aircraft were fully utilized. Our reconnaissance trip had been an on-again off-again arrangement and just the day before we'd been scheduled in with a flight to the Pole Station, not Byrd. But through sheer cooperation and a desire to make the most of everything, we'd been able to match our needs with those of the United States Antarctic Research Program. Our needed reconnaissance was happening.

Soon, I joined some of my colleagues and curled up on my down clothing in a quiet corner of the cargo area. I was now quite used to sleeping whenever and wherever. The weather dictated everything from field trips for stories and pictures to sending articles and pictures via the mail to New Zealand on the ever-changing flight schedules.

We had a textbook landing at Byrd, taxied to the terminal, opened the door, unloaded people and equipment, loaded people and equipment, shut the door taxied out to the runway and were once again airborne. It was almost like the stop had never happened. Try as I might, I could not find a good view to the outside for pictures. I thought, even with the station under ice, I might have got a shot of the radio masts or some people doing stuff!

Almost 6½ hours after leaving McMurdo, grinding over the white nothingness of the ice shelf and Rockefeller Glacier, we headed up the Skelton Glacier on the Hillary Coast south west of Scott Base to the Boomerang Range area where Peter and his team would do their work.

The Skelton Neve.

We squeezed around the windows to get a look and I imagined the New Zealanders with Edmund Hillary just 11 years earlier negotiating the glacier on Ferguson farm tractors laying the route and depots for the final leg of the TransAntarctic Expedition across the continent.

In a letter to Lois that night I would write, "We flew all round the area, and I can now say it was worth coming to Antarctica just to see the Skelton Glacier, Skelton Neve and the TransAntarctic Mountains. As we flew into the Skelton Neve-Boomerang Range-Taylor Valley area of interest, the pilot reduced altitude from 31,000 above sea level to be just 900 feet above ice level so we could get a good view of the terrain. There was a lot of turbulence and the bouncing, skewing, rising and falling motion sent a couple of the team members into heads-down, puke-out mode. My only regret is that you are not here to share it with me."

With nervous energy, I stood on the left side of the cockpit behind the pilot's seat. The flight deck had been reduced to essential personnel only, but even then, it seemed crowded with the four-person flight crew, Bill, our deputy leader, Peter, the field party leader, and me. It was tough to get my shots. Seeing my dilemma, the pilot very kindly pulled me round and made room for me to sit at his feet so I could shoot out the floor windows as we made low passes over possible landing sites. To steady myself I wedged my feet against the wall and the pilot's seat. I could not lean my arms or elbows on anything for fear of shake and vibration. I used both cameras, the Canon with a 135mm Soligor lens and the Rollei with an orange filter (to help with contrast), shooting at 500 at f8. The cameras bashed into my eye as I looked through the viewfinders. Adrenaline surged through me as I manhandled the cameras, one in each hand and tried to remain steady as the plane bounced and weaved and seesawed. I tried to pick up on the directions being shouted across the cockpit but "over there" or "get a shot of that" meant nothing as I could not see the pointing arms or see where they were looking. I just went with my gut instinct and kept on shooting. I'm just glad I didn't have to try and focus as well. The cameras were set to infinity. Oh, and yes, I reloaded the cameras as well, six rolls in the 12 frame Rollei and two 36-frame rolls in the Canon.

I'm not sure how long this furious exercise lasted, maybe half an hour or so. I know I committed the cardinal Antarctic sin of sweating. Couldn't help that with the combination of tension and excitement. I'd think I'd have perspired if I'd been sitting there naked with the window open. I could hear everyone on the flight deck breathe with relief as we climbed and turned for home. The team agreed it had been a worthwhile exercise and everyone nodded appreciation at the skill of the pilot and the efforts of the flight crew to make sure we got the needed information on surface conditions at possible landing sites.

The icing for this reconnaissance cake would now be up to me. Back at Scott Base I grabbed a can of beer and headed to my darkroom. The day before this mammoth run I had mixed up fresh batches of film developer and film fixer and washed and dried all the stainless steel developer spirals and tanks. I was ready. In the blacked out room, I laid out my films and by touch, wound the film into the spirals and popped them into the tanks, sealing the lids against light. From there I turned on the red safety light of the darkroom and counted down the minutes to develop, rinse, fix, wash and hang the negatives to dry. Relieved that I at least had images on the films, I ambled down to my office to start fleshing out my VUWAE story notes on my little green Hermes.

Instead, I looked at the growing pile of letters from Lois. Some even had scribbled notes on them for articles I'd either written or would be writing. A previously read blue aerogramme was on top and I flipped over to the back page...a P.S.:

"Yesterday Rachel and Hilary were in our bedroom and Hilary was tidying up the books on your table. Rachel said, "You can't put those books there as Daddy uses that plug." Hilary piped back, "He can't; he's at the South Pole." Rachel back again, "He can use it when he comes back."

I smiled, and reread the letter before I headed back to the darkroom to print out black and white proof sheets for the the team review. I was careful to maintain the sequence of the aircraft's low passes to gauge the height of the sastrugi and the effect this could have on a ski-equipped Herc making a greenfield landing. I really didn't like making proof sheets. The extremely dry humidity curled the film and printing paper. I did not have a proofing frame and had to make do with the enlarger easel and a piece of glass. After laying the paper down I then laid the strips of negative on top, keeping them in place with my fingers while I gently lowered the glass on top with my free hand. I switched on the enlarger light, exposed the paper for a few seconds, then dropped it into the developer and watched the magic of photography appear. From the developer tray the printing paper went through a rapid rinse and then into the fixer. Fortunately, my proofing operations went smoothly. I produced about a dozen sheets, fixed and washed them, and put them on the drier. It was now very late into the night and there wasn't anyone around but the duty folk. Exhausted, I dropped the prints on the leader's desk and went

VUWAE 13 ready to go into the field at the Boomerang Range region, 140 miles west of Scott Base. From left, standing: Mike Gorton, Peter Webb (leader), Barrie McKelvey, Alessio Ollier, Carlo Mauri. Barry Kohn in front.

to my bunk. I wrote to Lois, "Things seems to be getting busier here, work wise. But I'll just box on. I'll need to come home for a holiday!"

It was all good news the following morning. The gang had reviewed the proofs, confirmed what they needed to know and started on the final preparations for a put in within the next few days. Leader Robin summed up the effort later that day with a thump on the arm and a "good show, P-R-O, you did it again."

The VUWAE team headed to their exploration grounds on December 12. One motor toboggan had to be towed to the plane and would have its starter pulley repaired in the field. The other motor toboggan had trouble getting to the plane too but they were both loaded on with all the gear and supplies the group would need for about a month in the field. The 8000 feet above sea level landing for the Herc was not easy. It flew across the site knocking the tops off the sastrugi with its skis for about two to three miles before turning round and touching down on the now prepared strip. The engines were kept at high revs during the unloading and the giant plane used jet assisted take off equipment to lift clear of the ice.

Adelie penguins nesting at the Cape Royds rookery on Ross Island, north of Scott Base.

II.
ON THE HOME FRONT

"Hi there, Neighbour.
Guess you are getting settled in now and working hard."

So began a letter from Howard's wife Gin in early October. "Lois was thrilled with your phone call the other evening. She's keeping cheerful and is busy with her theatre artwork. At the moment, she's using our extending brush to wash the outside of your house."

But the main words I was delighted to hear from Gin were that Lois was determined to master the mysteries of driving our 1954 Citroen Big 15. "Tuesday afternoon Lois and I packed your car with the kids and the puppy and spent a hilarious half hour backing in and out of the basement garage," she wrote. "But don't worry, the car is still in one piece and the towbar proved to be tougher than the fence!"

A couple of days later, Lois began her story of learning to drive the car. Fortunately we had a very good friend in Lance Girling-Butcher, a journalist colleague, who took up the challenge and continued the driving lessons. "Lance has had me out all week," Lois wrote. "He asked me where I wanted to go to practice and I said the racecourse as that is where you took me to learn to ride the motor scooter; and I remembered those episodes! Everything went well, I got there ok and drove around and up and down. Then Lance persuaded me to go out on the motorway. I did it. I'm getting quite good at looking out the rear view mirror. Next day, we headed out and down around the port area and let the dogs go for a run along the beach and then more driving. It is exhausting. Then another day of driving and learning how to park. Bridget, Rachel and the two dogs came with us while Hilary was at school."

Lance was putting Lois through her paces in his Volkswagen Beetle. On one early practice run Hilary did not want to go with Lois as she thought her driving would be terrible. After a couple of streets, she informed Lois that it was all right. She was relieved and they continued on to the beach to give the dogs a run. As they headed homeward and right at the beach intersection with the main road "there was a helluva noise, like tins and glass being broken," Lois wrote. "The fan belt was in shreds so Lance and I put on a new one, adjusting it as well. My first fix-it lesson."

I was really encouraged for Lois when an early November letter announced that in Lance's opinion "if she learned her road code she should be able to get her licence!"

"I think I'll get an instructor to take me for one or two lessons" she wrote, "and he can book me an appointment and such. Hope I don't fail it."

A few days later I got back from a field trip to find the latest instalment in the License-for-Lois saga. "I managed to phone the driving school to get a couple of lessons. Oh dear, I've got butterflies already. They will phone me in about a week to 10 days."

At the end of November her letter reported that Lance and his wife Ally came to get her and Bridget for lunch. Lance drove down our very steep drive and told Lois she had to reverse up. " That'll be the day," she said. "I'll drive up frontways!" With a hiss and a roar they left the bottom and shot up the drive like a rocket. When she took the car down town, through the lights and into the gas station, the attendant chuckled, she said. "What are you laughing at? Does it show that much?"

"You get to know who's just learning," he said charmingly. "We've all had to go through it." And with that he cleaned the car window with a flourish and gave the car a little pat.

Lance's instruction continued into December, as he coached Lois on hills, downtown parking, and stopping at pedestrian crossings. He got Lois to phone the Transport Department and book in for her license test. "So, the written is on December 17 and if I pass that the practical is the following day," Lois wrote. The Marriotts kept the pressure on and made Lois drive the Citroen to her Mother's house. "I had the seat as far forward as it would go and a cushion under my bum."

"Darling, I will be so pleased to have you back to drive me all about. My drive with the instructor was not too bad. He said if I mastered holding the car with the clutch I would be OK. However, I was driving in his Morris Minor which is a little different to the Citroen and Lance's VW."

With the 17th come and gone, telegrams and phone calls, Lois had victory in her hands. A new letter arrived to put all of us at Scott Base in the picture. I read it aloud to everyone in the room after supper as they had all shared in Lois' reluctant pursuit to drive.

She wrote that it took her two hours after getting her license, to find the courage to take the Citroen out all by herself. "Finally I did, and now I have been out in it each afternoon, usually taking the Marriotts with us. I am glad I got my driver's licence." As you say, now I can suit myself and not have to rely on others for transport."

From this point on Lois relayed her motorised adventures and her letters told of the new experiences. I read snippets to my colleagues of how she negotiated corners, drove the two-wheeled farm tracks to our favourite beach, repaired a faulty windscreen wiper in the pouring rain and negotiating a herd of cows on the highway.

"Bloody hell," Taffy grinned after I had read that piece. "She doesn't need you now, mate. "We all laughed, and popped a beer to celebrate Lois' new skill.

December 19, 1968

My Darling Lois:

Got your letter yesterday and instead of lifting me it made me sad. The pictures. I wondered about it before, but after seeing you, I worry that you are doing too much. You seem to be packing a helluva lot in a short space of hours in a day. Please look after yourself. Perhaps now you have your license you might be a little more independent and be able to do things in your own time and not be governed by others. I know this is a terrible way to start a letter but I think it best to say it. At least you'll get this letter after Christmas so it won't ruin that for you!

I'm feeling a bit whacked today as though someone walloped me with a sledgehammer. Most of the others have taken off to see the icebreakers about seven miles out. But I thought I would stay and write to you and get an early night. I notice you have had your hair cut. What color is it now? Back to original or still coffee?

It is nice to know that I have not been falling down on the job. Robin is quite happy with what I've sent out and I got a letter from NZPA yesterday to say that output and quality has been good.

This week has been the warmest with temps just a little below zero. On Tuesday it actually got within a point of zero Celsius. So at home that would barely register frost level at 32°F. However, the wind does make it a lot colder.

I do not know what I am going to do with our life. At the moment I just feel like coming back to New Plymouth and quietly settling down. By coming here, I unloaded all our money troubles on you. But lately you have said nothing about them. How are they going? How's the car? How much do we have in the bank? With the reduced income I suppose you are finding it a bit of a struggle. If I come back to the newspaper, at least we'll be on a good pay, as good as we'd get anywhere in New Zealand. At least it goes further.

I promise the next letter will be a happier one!

Your man at Scott,

G

December 22

Hello My Honey:

I felt dreadful about posting that last letter. I went to get it back from the mail only to find a plane went out today much to everyone's surprise. Our postal team is really great at making sure our mail gets out.

I aim to make this a good letter, happy and bright for my co-driver. I suppose you are now zooming all over town. I am now writing this letter from a very snug and comfortable bunk after a very torrid Sunday. I got up at 8 am and made myself some breakfast with John Newman and Noel Wilson before heading to my desk to write a story. I found I wasn't in the mood, so I shuffled up to the darkroom to process a couple of rolls of film I shot at a party we had last night for some officers from over the hill. Awful party. Not our usual style at all. Nobody seemed to get a spark.

It was my shower night tonight and I feel all lovely and clean. I've got my Johnson's baby powder on. It is provided in very large quantities here and used by all in lieu of water! So here I am scrubbed, clean-smelling, in fresh pyjamas, and warm.

I was up at 5:30 am today to finish my story on Les Quartermain, a wonderful fellow who has become a firm friend of us all. His knowledge of all things Antarctic is amazing. I sell one of his books *South to the Pole*, a magnificent tome which he published last year.

After breakfast, I did the dishes and other mouse duties. I was supposed to go out to Vanda at 8am to get some pictures for Antarctic Division use but the weather is dirty. Have not yet heard if the flight is scheduled for this afternoon. Consequently I'd like to get this wee letter completed in case of panic when I get back from Vanda. My typing is a bit slow right now. I bashed the end of my finger under a chunk of ice while I was filling one of the melters. It is quite sore but not damaged.

I have put three of your photos up in my bedroom and I have one I mounted on a card here at my desk. I mention this now as I just looked up and saw you smiling at me.

I called NZPA today about my story on the seals. I dated it November 28. It was used "somewhere." Perhaps you might check at the Library reading room. It would have been published in the first week or so of this month.

Well, lunch is on now. Be back soon. We had lovely roast lamb, mint sauce, gravy, plastic spuds and mixed veg. This is the meal I like down here. It is the best meat on the base.

I have to get ready now. The helicopter is due in half an hour. I'll gamble and see if I can complete this when I return. Its 5:15 pm and I've just arrived back. It was a great trip up the valley. Seemed just like landing on the moon. It is all brown with the Onyx River braiding down the middle of it. On the way back we came down the Victoria Valley and that is different again. The valley floor is more or less just sand and grit compared with the Wright Valley that is bouldery.

Sorry dear, but I have to end here and rush and get my other story and pictures packaged up for the mail. The flight is about an hour earlier than expected. Cheerio for now. I'm really looking forward to the plane with mail tonight and to phoning you on Christmas Day. It's exciting.

Your base shutterbug,

G

When the sun shines take advantage...and on this one day it was warm enough to find a place in the lee of a hut for afternoon tea and coffee.

The US icebreaker Glacier forces a channel through the McMurdo sea ice towards Hut Point so the resupply ships can get in to both the US McMurdo Station and New Zealand's Scott Base.

12.

FOR LOVE OF AN ICY PLACE

One of the first photo assignments I had after arriving at Scott Base was getting a picture of Robin and his US counterparts unveiling a plaque at the out-of-bounds area around Scott's historic Discovery Hut, which overlooked McMurdo Sound on the northern perimeter of the US McMurdo station. The sign read:

"Hut Point was the site of Captain R.F. Scott's National Antarctic Expedition 1901-1904, the first expedition to penetrate the heart of the Antarctic. Discovery [the ship] was iced in close to the shore on the southern side of this point and was the home of the expedition. This hut was constantly in use and in later years was the main staging post for parties striking south towards the Pole in 1907-1909, 1910-1913, and 1914-1917. This is historic ground."

The huge laser etched wooden sign had been screwed to the side of the hut that was surrounded by a heavy chain boundary fence. It was out of bounds to all personnel at the US station. The NZARP team, as conservators, and guests could only visit with our leader's approval. I wished I could get closer and take a peek inside. But I was on deadline and had to get back to base to process the film and get prints and captions for NZPA into that day's mail-bag.

My chance to get inside the hut came just a couple of weeks later when Antarctic Division asked for a new series of colour slides of the Discovery Hut. I'm not sure what I expected, but I was totally surprised at the heavy snow accumulation and the bleak, soot black interior. The four foot wide

205

verandah all round gave the hut the appearance of being much larger than its meagre 36 square feet. Scott had brought the building from Australia but it was not used for accommodation that first expedition, serving only as a storage facility. The hut's real use came during subsequent expeditions as a relief staging point and refuge hut. The attempt to insulate or wind proof the hut with sacking and felt nailed to the thin boards was still evident. The crude lining was now blackened and tattered. It was a dingy, dirty, dark smelly hole that bore the scars of time, neglect and hardship. I wondered at the voices within the walls as I looked around and photographed the battered shelving still holding biscuit tins, tins of oatmeal and even tins of mutton. Dotted along the shelves were the blue-green vertigris crusted remnants of oil lamps and spirit stoves. In the centre of the building was the crudest of stoves with a rusted kettle still in place. Sledging gear leaned against the wall in the entryway. The hut reminded me of a burned-out house after the firefighters had gone. I wanted to visit the other huts further north on Ross Island.

Scott's 1901-1904 Discovery Hut in 1968, with the shelves remaining stacked as they were left.

A few days before Christmas saw me with three companions scurrying north across the fast ice (sea ice frozen along the coastline) of Erebus Bay to Cape Evans, the site of Captain Scott's 1910-1912 expedition hut. It was a rare day of gloriously fine weather, clear blue sky, minus something temperatures of the who's counting variety and wind, wind, wind. In spite of the great conditions, I was (and I think we all were) on the tense, nervous side of normal as it was well into the summer season to be using the bay as a highway. The so-called fast ice varies anything up to two meters in thickness between summer and winter. Field reports indicated that the ice was stable, but that did not shake the apprehension and tension from my bones. Ice, sitting on the sea, is just that...ice sitting on deep, cobalt blue freezing sea. And there's no getting away from that.

US and New Zealand erect an information board at Scott's 1902 Discovery Hut. In 1968 a chain barrier was erected around the hut, declaring it out of bounds. Robin Foubister, leader at Scott Base, is at left.

Chippy John Newman repairs a door to a university field hut at Cape Evans.

Black lava rock hills and glacier preclude any practical overland route from Scott Base to the cape. Even the daring men of yesteryear avoided it as it was very tough terrain to man-haul a sled across. In the early 1900s, they waited at Hut Point (Scott's first hut) for the fast ice to be firm enough for travel. Once we rounded Hut Point, it was more or less a straight 25 mile run, skirting the Erebus Glacier Tongue to the east and passing by Razorback Island to the west and round the tip of Cape Evans to the hut, just off the beach.

I had a double brief for the day trip organized primarily for me to get New Zealand's "Mr Antarctic", Les Quartermain, to revisit the historic Cape Evans hut. As well, the Antarctic Division library in Wellington wanted some updated colour slides. A walking Antarctic encyclopedia, Les had already visited Scott Base on two occasions, but, now we'd managed to arrange his first trip to the South Pole. His other interests were to revisit the historic huts and memorials at Cape Royds, Cape Evans and Hut Point. Throughout his 12-day tour, the 73-year-old former schoolmaster spent many hours sharing his vast knowledge of Antarctica's pioneer exploration with Americans, Japanese and New Zealanders at Scott Base. He was the author of several booklets including his new book on the early history of the Ross Sea sector, the first full scale account of any one sector of the Antarctic. It was a book I sold to many of our visitors on behalf of Oxford University Press.

Les' interest in the Antarctic started as a high school boy in Christchurch when he watched the return of Shackleton's Nimrod to Port Lyttelton in

1909, heard a lecture given by Shackleton and watched the departure of Scott's last expedition in 1910. The man had history. He was a foundation member and a former president of the New Zealand Antarctic Society, former editor of their quarterly magazine, and author of several booklets for schools. His current project, and part of the reason for his third visit south, was to complete his history of New Zealand's association with Antarctic research and exploration.

The remains of a stove in Scott's Discovery Hut . Although not built for accommodation, sledging parties used this hut in subsequent expeditions en route from bases at Cape Royds (Shackleton) and Cape Evans (Scott).

Any field trip from Scott Base called for a minimum of four people, so we took advantage of Deputy Leader Bill Lucy's availability as navigator and for John "Chippy" Newman to go up and fix a door on a university field hut. Diminutive Les was alive with anticipation, and his excitement at being where he was and where he was going flowed to us. He had stories to tell of his many years and contacts made in his detailed research for his books.

I was sure the conviviality and friendship in our little red Snow-Trac would rank up there among the memorables of my time on the ice. There was the usual bit of travel tension at the start of any field trip, especially one over the sea, but I was happy for this day and the opportunity to travel over land once more in contrast to the bounce in-bounce out option of a helicopter trip.

Out from Scott Base the going was smooth, but once into the open bay the surface took on a different texture as our little rubber-tracked heater-less snow machine negotiated the sastrugi. Bill constantly sought smoother passage to provide his passengers with a degree of comfort.

With a panoramic view across McMurdo Sound, I could see for miles with the TransAntarctic Range on the mainland to the west and the smoking Mt Erebus on Ross Island to the east. As I settled into the rhythm of the

The stables at Scott's Cape Evans Hut.

Adelie penguins at Cape Royds.

journey, the tracks clattering smoothly around the drive wheels beside me, I relaxed.

Les chattered non-stop in his boyish excitement telling us about the adventurers of yesteryear who took a couple of days and more to battle across this same terrain we'd cross in a couple of hours. He talked about their primitive gear, man hauling sledges and poor physical condition after being weeks out in the bleak and inhospitable Antarctic environment. We were marveling at their strength of character and determination, when we came upon our first group of Adelie penguins.

Now I was squiggling round to get a view, hauling out cameras, banging poor Bill on the shoulder with a "Look, Look!" We were excited puppies. I knew this little scene would be front and centre on the next mail home.

Carefully, Bill pointed the Snow-Trac towards the tuxedo wearing quintet tobogganing along on their bellies. They had quite a turn of speed but when they spied us coming up to them they stopped, stood up, and stared at these out of uniform tourists on their patch. They stood there unafraid, posing as if in some old-fashioned grandfatherly formal family photo. After assessing the situation they nodded to each other, choosing to ignore us. They turned and waddled off about their business. We headed toward them at a safe distance so we could get close enough to capture some good shots on colour slide film. That's when they decided enough was enough and started yelling at us and flapping their wings in a watch-it-bucko attitude. Looking back as we headed off, our lasting view was to see the Adelies in a huddle, head to head, their flippers outstretched, no doubt muttering about the nosy Kiwi tourists interrupting their day.

A bit further along we came across a pair of US icebreakers crushing their way south to make a safe channel for the supply ships to reach the wharf at McMurdo. Again, another unforgettable photo op. We left the confines of the Snow-Trac and got to within 20 feet or so of the giant white ships as they

The boot rack at Shackleton's Cape Royds hut.

rode up, bow first, on the sea ice, crushing it under their weight and pushing huge chunks to the side of their channel. From sea level I watched the big ship back up and then power forward in an awesome surging impact of steel, ice and deep blue ocean, rising, flooding and subsiding.

As we pushed on to Cape Evans, I was glad we had Bill with us as we faced the hazards of open leads in the sea ice. In other words cracks! And they were more obvious and wider than what we had experienced in October on the tractor train. We were all eyes ahead and to the side. At times we had to travel parallel to the crack until we found a safe place to cross. While we were content to sit in the university hut and enjoy a brew and food, Les was truly in his element. The area around the hut had not changed since the remainder of Scott's party left in 1912.

In my letter to Lois that weekend, I wrote, "Les took off like a gnome and disappeared into Scott's hut." It was almost a decade since his last visit when he and others did a magnificent restoration job on the hut that had remained deserted and filled with ice and snow for some 40 years.

He emerged some 20 minutes later grinning from ear to ear. He didn't say much, just enjoyed the moment as we fed him and gave him a steaming mug of hot chocolate.

"Nothing has changed," he said finally. "It is pretty well as we left it. Seeing it again has certainly made my trip."

For the next two hours, Les wandered around inside the hut. His mind traveled back 43 years

Herbert Ponting's darkroom at the Cape Evans hut.

when Captain Scott used the hut as a winter base before setting out on his final journey to the South Pole. It was education time for me as Les, ever the teacher, explained where each member of Scott's expedition slept and where they had worked. He told me in great detail the task his restoration group had in cleaning out the hut and what they had found. He rummaged around through the old broken-open packing cases and the burst straw bales outside the hut before heading off for a walk up the hill to the memorial cross from Shackleton's Imperial TransAntarctic Expedition of 1914-17 Ross Sea Party deaths of Captain A. McIntosh, V.G.Hayward and Reverend A Spencer Smith.

When he came back down the hill I could see from the glint in his eyes that Les was in seventh heaven. Antarctica had been his life's interest and he was here again. He was absolutely ecstatic when I showed him what I had uncovered while he had been up the hill, a pile of pony snowshoes in an exposed pond near the entrance to the hut. At his bidding, I fished out a manhauling sledge that had lain there unknown for half a century. As far as we could tell, this was the first time the ice-covered pond had melted.

"This is remarkable," Les exclaimed as he wandered around picking up the snowshoes as I fished them out of the pond with the aid of a long pole. "I didn't know this pond existed." From what I could gather from Les, the pond

The bunk I believe was inhabited by Captain Oates of Scott's South Pole expedition. Heroically, the severely frostbitten and ill Oates walked out of the tent to his death on the return journey from the pole.

had lain concealed under ice and snow for more than half a century. Bill, a six summers man in the Antarctic, confirmed he had not known of the pond despite his many visits to the hut.

Like many a kid out tadpoling in New Zealand, I fell in the pond, not so much right in but I did get my feet wet. Not a good thing to do in this frozen environment. Hearing all the excitement, Bill and Chippy joined us. We fished out 16 snowshoes. Prior to this day, only eight such shoes had ever been located; six of them were in the hut and two were in the Antarctic Division's headquarters in Wellington, New Zealand.

According to Les, the sledge was a rare find. It was complete even down to the manhauling harness. The sledge was in better condition than the one housed in the hut. I photographed the sledge and the new finds and we put them in the hut for historical safekeeping. Les would advise the NZ Antarctic Society and Robin would log the find with the Antarctic Division.

With Les due to leave next day, we figured it was time to turn for home. I was feeling very uncomfortable with my extremely cold wet feet, but they'd be fine now I was off the ice and snow and inside the vehicle. At least we did not have to worry about nightfall. I nodded off very easily in the 24-hour daylight as we trundled homeward. Les captured a bit of shuteye too, in between chattering about the marvellous time we'd given him and the discoveries we'd made. Napping in the Snow-Trac, I forgot about my cold feet and the slight undulation of the sea ice!

Graeme Connell with Les Quartermain examine pony snow shoes found in a melt pond outside the Cape Evans hut.

A month later I received a letter from Les. He had contacted Dick Richards, a veteran of the Shackleton Ross Sea Party, who wanted more information about the sledge we had located. Les wanted to bring some context to the sledge and suggested it could have been one "brought from the south", a remnant of Scott's trek to the pole. Les rounded out his letter asking me to express his gratitude to all at Scott Base for making his visit so memorable and for the "grand time you gave me."

It was now too late in the season for me to get back to Cape Evans over land so I sent Les what pictures I had. By chance I did get back to Cape Evans on February 7 on a quick helicopter trip. The long light of summer was rapidly departing and we were experiencing almost regular day and night. As a result chopper time was hard to come by. Somehow I was able to hitch a quick ride up and back to get the picture of the sled for Les. My seat on the chopper was probably the result of our good relations with our American colleagues and of Robin's never-ending ability to find a way.

The sea ice to that point of McMurdo Sound had now broken out and all you could see was the brilliant blue of the ocean. Barely a month before, I had been driving across that part of the Ross Sea in what amounted to a customized Volkswagen!

Excitement For Antarctic Expert

SCOTT BASE.

By Graeme Cornell, N.Z.P.A. special correspondent.

A life-long interest in Antarctica and the men who have sought to conquer the icy wastes reached a climax recently for Mr Les B. Quartermain, Wellington, author and former schoolmaster, when he visited Scott's famous last expedition hut at Cape Evans, about 20 miles north of Scott Base.

During his 12-day stay in the Antarctic 73-year-old Mr Quartermain:

● Became one of the oldest people to visit the South Pole Amundsen-Scott Station.

● Stayed overnight at the historic Cape Royds Hut erected by Shackelton in 1908.

● Made day visits to Scott's huts at Cape Evans (1910-1913) and Hut Point, McMurdo (1901-1904).

● Made two ascents of nearby Observation Hill to inspect the memorial cross to Scott and the members of his ill-fated expedition to the Pole.

● Spent hours sharing his vast knowledge of Antartica's exploration with Americans, Japanese and New Zealanders at Scott Base.

This is the third visit Mr Quartermain has made to Antarctica. His first trip came "out of the blue" in 1957 when he was invited by Operation Deep Freeze to join a Press party. He was then editor of the New Zealand Antarctic Society's quarterly, Antarctic.

It was the second visit in 1960-61 that gave Mr Quartermain his greatest opportunity. He was selected to lead a small expedition from New Zealand to clean out and restore the Cape Royds and Cape Evans

Seeing it again has certainly made my trip. . . ." he said.

For the following two hours he walked around inside the hut. His mind went back 43 years when Scott used the hut as a winter base before setting out on his final journey. Mr Quartermain explained where each man had slept, where they had worked, the task of cleaning out the hut, what the party had found. Two - thirds of the, hut had been filled with snow. The restoration of both huts had taken two months.

Outside, Mr Quartermain fossicked through the old, broken - open packing cases, the burnt bales of straw and strode to the top of a nearby hill to inspect a memorial cross and plaque.

The greatest excitement of his brief visit came when I located some pony snow shoes in an exposed pond near the entrance to the huts. I also fished out a man-hauling sledge that had laid there unknown for almost half a century. "This is remarkable." Mr Quartermain said as he wandered around the pond picking up the snow shoes as I fished them out with the aid of a long pole. "I didn't know this pond was here."

Sixteen snow shoes were found and placed inside the hut for safe-keeping. Until that day only eight snow shoes were in existence — six inside the hut and two in the Antarctic Division's office in Wellington.

◆ ◆ ◆

The sledge was a rare find. It was complete, even down to the man-hauling harness. Mr Lucy, who has visited the hut many times before in his six summers on the ice, said he had not known of its exist-

Lytteiton in 1909, heard a lecture given by Shackelton and watched the departure of Scott's last expedition in 1910.

◆ ◆ ◆

Mr Quartermain is a foundation member of the New Zealand Antarctic Society and was president in 1957-59. He was founder of the society's quarterly, "Antarctic". He retired from editorship this year after producing 50 issues.

He has written several booklets for schools, and in 1967 his major work, "South to the Pole" was published. The book dealt with the early history of the Ross Dependency.

Mr Quartermain is at present working on a complete history of New Zealand's association with Antarctic research and exploration. It is scheduled to be published next year by the Government printer.

The Cape Evans memorial cross erected for three men (Hayward, McIntosh and Spencer-Smith) lost during Shackleton's 1914-16 Imperial TransAntarctic Expedition.

Scott Base leader Robin Foubister. I sent a similar picture to the NZBC's Listener magazine for inclusion in a Christmas feature on New Zealanders abroad.

13.
CHRISTMAS ON ICE

A four-mile hike over the hill to McMurdo for Christmas Eve festivities seemed easy knowing good food and conviviality awaited us. At "refreshment spots" there, we wined and dined like kings and taught our hosts how to talk like Kiwis. Somehow it did not matter that the southern summer night came with a wind chill of over -25°C. A young Texan in the Sergeants' Mess introduced me to his BBQ — a backyard charcoal grill located in the snow outside an escape hatch, roughly a 30-inch by 30-inch trapdoor in the side of the hut. A true steak artisan, he'd open the hatch, check his T-bones and ribeyes, and close it back up again. The smoke from the grill stayed outside. I just hoped nobody would have to use the hatch and noted that there were at least two others in the room. Over at the United States Antarctic Research Program headquarters (USARP), Byrd Station ice cooled our Scotch. Here scientists were drilling thousands of feet through the ice cap into landmass below sea level. Reading rings on the core of ice in their drilling hole, they figured it to be pre-Christ. And here it was in our glasses, BC ice popping and crackling around the crowded room, each chunk lasting almost the entire night.

Close to midnight, we wished our American friends a merry Christmas, and a group of us made our way to a large Quonset building for a special all-comers church service. There was remembrance of the US and New Zealand military in Vietnam. Though the fullness of what Christmas is really all about had not settled with me at that point in my life's journey, in view of my Wright Valley experience some weeks earlier, I willingly joined others in the service. In the dimly lit hall, I added my Flat D to the bass, baritones

The Scott Base group gathered for Christmas dinner.

and tenors of hundreds of men singing "Silent Night". This most unusual of Christmas venues did begin my slow spiritual awakening.

I was among the early risers on Christmas Day and sat quietly in the Mess hut, troubled. Chippy came in and sat beside me.

"Think I have to go to the hospital this morning," he confided. "I'm pissing blood."

"Me too," I replied. "I'm gonna wait and see what happens next time I have to go."

Normally very talkative, we didn't say much else. A couple of the others came in and sat down. The room did not have the feel of a Christmassy morning.

Robin arrived grinning ear-to-ear. "Everyone feeling ok this morning? Merry Christmas!"

Silence. He'd been talking with the McMurdo doc, Roger Case, and learned several people had turned up at the Sick Bay early that morning because they were peeing red. The medics had spiked the Christmas punch with a harmless medical dye. Now we switched to relief and laughter and got on with Christmas. Like others, I'd booked a call to Lois and the girls and was first on the schedule. Those of us with young children missed the patter of little feet at 3 am.

The room seemed full of people, some of whom were sitting on or at my desk. I locked the door to the telephone booth.

Ice outside in the snow melter sledge made a handy dandy wine cooler.

Being so far away hit me, and tears rolled out as I greeted Lois. I didn't get time to say much to my girls or to dry my tears. As soon as I emerged from the booth, the first to spot me was Chippy.

" 'ullo," he says in his best London accent. "Whatchoo bin do-een? Look at 'im, look at 'im...ha ha ha." The whole room followed his cackling laughter as there were at least half a dozen others waiting in line. I went over to the window and stared out at the frozen landscape. Robin started laughing and said it must be because I had daughters to his four sons, but as it turned out, I wasn't the only man with moist eyes. I just happened to be first.

Now we all had work to do before our Christmas dinner in the evening. Still, the day was special. Visiting with the US science group the night before had got us pretty excited about the Apollo 8 astronauts, the first men to ever fly right round the moon. As they orbited, they read the Bible creation story. We got to hear bits of this during the day when our Post Office techies rigged up a speaker to their radio system.

Throughout the day many of us connected with pals out in the field. I'd sent a story through to NZPA and NZBC outlining the whereabouts of field groups and how they would celebrate the special day, far from base and far from home. There were 18 New Zealanders, four Italians and two Japanese in the field. These 24 men were well spread in the McMurdo region, four

Summer Deputy Leader and Vanda Station leader Bill Lucy leads the group in a toast to all the folk out in the field.

Victoria University geologists and two Italian alpinists in the Boomerang Range region, four Canterbury University biologists at Cape Bird on the northern tip of Ross Island, 10, including two Japanese scientists, gathered at the new Vanda station. The loneliest pairs were two Italians exploring in the Upper Wright Valley and two New Zealand field assistants working down on the Wilson Piedmont glacier. In a letter from Buckingham Palace, the Queen sent her best wishes for Christmas and this, as well as other greetings were conveyed by radio to the far flung field parties. For these adventurers, Christmas dinner was little different from the usual field rations cooked over a kerosene (paraffin) stove. All field groups did get a bottle of New Zealand wine. The Vanda group had their rations supplemented by a Christmas cake, two chickens and a goody box of nuts, biscuits, potato chips and candy. The Cape Bird team at least had a sip of wine, but they missed their Christmas mail when operational difficulties forced a planned helicopter trip from an icebreaker to be abandoned.

For the 18 of us at the base, cook Geoff Gill served up a scrumptious dinner of roast chicken, spring lamb, and pork followed by steamed Christmas pudding.

We toasted the field parties, the biologists, geologists, meteorologists, glaciologists, and physicists, and our wives, sweethearts and families. My story of the festive day received a wide play in the New Zealand media.

We wish you a merry Christmas.

A real family Christmas card from home.

December 26, Boxing Day

Hullo Pet:

Sorry about all that watery stuff yesterday. Don't really know what happened except I was glad to be talking with you. I'm sad I could not say much to the girls either. They must have thought it pretty strange. Chippy spotted my wet face when I got out of the booth and let everybody know. But when he finished his call home, he too was a bit wet eyed.

I hope you might have seen on tv or read in the newspaper my NZPA report on the North-South rugby match on Christmas Eve. Well, I got thumped a real beauty. I am now hobbling around with sprung ribs. Nothing serious. Just darn painful. I visited Roger (the doctor at McMurdo) today, for an examination, and he gave me some pills to take away the pain and help the joints knit together. I am not much use around here for at least three weeks. I can't lift anything heavy. It happened when I tackled a fellow on the goal line. He lost the ball and I dived on it to force it. As I dived, 210 pounds of Foubister dived at the same time. I grabbed the ball and tucked it under me right into the sternum and he landed on top. Crunch! Nothing broken though. The ribs apparently just sprung out of the joints and popped right back in.

I'm blowed if I know what I am going to do when I finish here. Perhaps we should move to a newspaper elsewhere in New Zealand. Think about it, love, and let me know your thoughts. Any nibbles on the house? How is the Citroen doing? And what about the bills? Are they mounting up?

It is 5 pm now and I have an hour before the mail closes and I still have to go over to McMurdo. I suppose I could leave it till tomorrow but there just might be a picture available of me playing Rugby that I can send with this letter. So I am going. Like yesterday I was up and about at 6 am today so that I could get eight films processed. And because I did this I can send you something, like the group at Christmas. This afternoon I finally got thoroughly brassed off with the amount of light getting into the darkroom so I spent about three hours doing a blackout job. It is now pretty good. That is all for now my love. I might have time to add a footnote when I get back from McMurdo. Enjoy the pictures

Him with the red beard

G

Mid January

Hullo My Darling,

I'm certainly looking forward to your homecoming. I think I'll kidnap you and lock you in a little room. We do love your letters and pictures, so keep them coming. I just loved your phone call on my birthday. That was truly terrific. Don't buy me a present, the phone call was enough!

Gin and Howard are home so I won't be quite so lonely. Tomorrow I have to bake a cake for one of my art ladies whose daughter is having a birthday. She said she cannot seem to bake a cake in the gas oven so I stepped in and offered. Gratefully received. Gin brought me 20 pounds of plums so I canned those and then we drove over to Mother's for lunch. I spent a couple of hours digging prickles out of Dad's fingers. They were quite holey by the time I had finished. Father said: "There's nobody can get them out like you do, dear. Another thing you do that nobody else can do is cut our daughters' poached eggs on toast. I invariably get, "Daddy doesn't do it this way." So, my love you are not forgotten.

Hilary has lost her top front tooth and she looks a character. She wiggled away at it herself. She let me tug at it twice. It would not come for me and then lying in bed, kinda grating out it came. The tooth fairy gave her 15 cents for it because she had done it herself.

This afternoon, after growling at our two eldest daughters, I asked them what was the matter and why was it necessary for me to growl. Both at the same time they chorused: "It must be because we miss Daddy."

"I do hope he comes home soon," says Rachel.

"I want him to look after all of us again," Hilary added.

Kisses from all of us,

Tuppy

January 5, 1969

Sweetheart

Visited family and friends over New Year. Father greased the car for us and noted the tyres were a bit worn but would probably be ok till you get home.

We saw your rugby match on the television. I was so pleased I hadn't missed you and so were the girls. We just had time to remark, "There, he is, Daddy ..." and you were gone. It was so nice to see you."

I don't know what we should do when you get back either. I know you say you'd dearly love a rest, but do you honestly want to stay in New Plymouth? I don't think you would like it back at the newspaper but if you think you can handle it by all means we'll stay here. It is a bit of a pickle isn't it? I'm all for you and behind whatever decision you make.

Just before Christmas, I had a couple from Wellington come look at the house. They will be transferred to New Plymouth sometime in the New Year. They liked everything about our house. His parting words were, "Hope it isn't sold when we come back." So I don't know whether to get my hopes up as I don't want to get all eager. They have to sell their house in Wellington.

All my love

Tuppy

New Zealand units:

Star, Dunedin 27/12/68

QUIET CHRISTMAS IN ANTARCTIC SCENE

(N.Z.P.A.—Special Correspondent.)

SCOTT BASE.—Eighteen New Zealanders, four Italians and two Japanese spent a quiet Christmas yesterday in the field. For them, Christmas dinner was little different from the usual field rations cooked over a primus stove.

The 24 men are working under the New Zealand Antarctic research programme—four Victoria University geologists and two Italian alpinists in the Boomerang Range region, four Canterbury biologists at Cape Bird on the northern tip of Ross Island, six New Zealanders, two geologists and two Japanese scientists at Vanda station in the Wright Valley and two Italians in the Upper Wright Valley.

The loneliest Christmas was spent by Messrs Bruce Brookes, Christchurch, and Derek Cordes, Rangiora, who are working on the Wilson Piedmont Glacier.

HOME WINE

But for all the field parties, Christmas was brightened by a bottle of New Zealand wine.

Two field groups working in the Wright Valley converged on Vanda station on Wednesday. The six men—Messrs Tom Broxley, Ron Craig and Lester Tomlinson, (Wellington), Alister Ayres and Simon Cutfield (Auckland), and Charles Hughes (Dunedin), were joined by Messrs Burton Murrell, Victoria University, and Russell Blong, Sydney University.

two Japanese scientists, Dr T. Torii and D. N. Yamagata.

Their rations were supplemented by a Christmas cake, two chickens and a "goody box" of nuts, biscuits, potato chips and sweets, sent out from Scott Base.

The Wilson Piedmont Glacier party enjoyed similar treats.

At Cape Bird, Dr Euan Young and Messrs Tony Harrison, Trevor Crosby and Eric Spurr could only supplement their rations with New Zealand wine.

DELAYS

They did not even receive their Christmas mail. The mail was to have been delivered by helicopter from a United States ice breaker but operational difficulties forced the flight to be cancelled.

The Victoria University group comprising Dr P. N. Webb, Dr B. C. McKelvey and Messrs Barry Kohn and Mike Gorton, and two Italian alpinists, Messrs Carlo Mauri and Alessio Ollier, celebrated Christmas on Christmas Eve as Wednesday was to be spent working.

Two other Italian guests of New Zealand, Dr Marcello

Manzoni and Mr Ignasio Piussi, also had an isolated Christmas. They are working in the Upper Wright Valley, far from any Antarctic civilisation.

At Scott Base 16 New Zealanders and two Englishmen celebrated in the comfort of a warm, decorated base. Most telephoned their wives and families in New Zealand during the day.

TRADITIONAL

Christmas dinner was the traditional fare of roast chicken, lamb and pork, followed by steamed Christmas pudding.

Greetings were received from all over New Zealand and included telegrams from scientific services represented in the New Zealand research programme.

In a letter from Buckingham Palace, the Queen sent her best wishes for Christmas and greetings were also received from two other New Zealand outposts, at Campbell and Raoul Islands.

The greetings were all conveyed to the far-flung field parties by radio.

A telegram from the Italian Ambassador in New Zealand was passed on to the Italians working with the New Zealand parties at the Scott Base dinner.

Toasts were made to the Queen, the New Zealand Antarctic research programme 1968-69 (by the leader, Mr Robin Foubister, Christchurch), field parties (by the deputy leader, Mr Bill Lucy, Timaru), scientific disciplines (by the scientific leader, Mr Peter Lennard, Christchurch), to wives and sweethearts (by the base maintenance officer, Mr John Newman, Dunedin), and to O.A.E.'s (old Antarctic explorers), (by the base electrician Mr Chris Rickards, Taupo).

Radio operator Bob Hancock at his Morse key sending a message to the New Zealand Post Office.

14.
OUT OF THIS WORLD

dah-dah-dah dit-dit-dah dah dah-dah-dah dit-dit-dah-dit dah dit-dit-dit-dit dit-dit dit-dit-dit dit-dah-dah dah-dah-dah dit-dah-dit dit-dah-dit-dit dah-dit-dit

That's the rhythm of my neighbour at work in the adjoining office, radio operator Bob Hancock tapping out my news release on his Morse key, the miracle code sending letters, words and story to the New Zealand Press Association via New Zealand Post. It was the story of New Zealand's Governor-General, Sir Arthur Porritt, lifting a lever, setting a wind generator into operation and declaring the country's first Antarctic mainland winter station officially ready for business, the primary effort of the 1968-69 New Zealand Antarctic Research Programme summer.

It represented maybe two of 150 summer days that I spent writing about the everyday stuff of an expedition in the inhospitable environment of Antarctica. Behind those two days lay weeks of ingenuity, initiative, effort, sweat, risk and fears. A lot of guys had gathered and relocated materials and supplies into the Wright Dry Valley overlooking Lake Vanda and reassembled huts, winterized them and got everything into shape to provide a measure of comfort for five brave souls remote and isolated over the long polar night. Theirs would be a first in that haunting rock and grit moonscape. It had been a mammoth task for members of our team with visiting help from New Zealand.

We started preparing for Sir Arthur's visit with his two sons Jonathan and Jeremy, and his aide-de-camp Captain James Innes about a week or so

New Zealand's Governor General Sir Arthur Porritt greets an Adelie penguin at the Cape Royds rookery.

before his arrival on January 8. Our preparations were very basic: make sure the place is clean and tidy and have our desks organized and make sure all the books and magazines in the mess hut were put away.

Under the theme of international co-operation, our American neighbours were joint hosts and on January 7 took the Vice Regal party to the South Pole. I was scheduled on the trip but suggested my seat should go to someone else as I had work coming out my ears with developing films and writing media reports. Spending what would be eight hours or so in a Hercules was a time expense I could ill afford. Besides I had already been to the Pole. Doug Spence, our storeman, was given the seat, and I was thrilled as he was a very hardworking fellow, working most of the time on his own in the cold confines of the hangar. As a result, he did not get out and about much and this would be his only chance to get to the Pole. I gave him the base Rolleiflex and asked him to get a picture of the Gov's younger son Jeremy doing a handstand at the Pole. He did and I sent that to the media under the headline, he "Held the World in His Hands".

I had a busy night ahead of me when the group arrived back from the Pole. I basically went where His Excellency went, shooting film for posterity and the Antarctic Division files. Media coverage included raising the Vice Regal flag at Scott Base and the South Pole picture. After cocktails in the Post Office/Communications hut (my office in one corner) we had a slap up dinner that featured excellent New Zealand roast lamb. Our mess hut was not the greatest of places for a vice regal dinner, but I'll bet it was one of the coziest, relaxed and friendliest places His Excellency had ever dined in. Our chrome, vinyl and Formica furniture in undraped painted walls would contrast sharply with the brocade and velvet, shining silver and crystal of

The Vice Regal party outside the Shackleton Cape Royds historic hut. From left, Master Jonathon Porritt, Sir Arthur, Master Jeremy Porritt, Captain James Innes, Sir Arthur's aide-de-camp.

vice regal milieu. In that very crowded, convivial room, we could count Italian, American, Japanese, British, New Zealand, Australian, Canadian and Russian guests as well the Kiwi team members.

After dinner everyone felt pretty rosy and Chris Rickards, our base electrician, hit the piano for a raucous singalong. Assistant surveyor Alister Ayres added to the mix guiding the happy diners through some good old campfire songs including the Boy Scout standard "Back to Gilwell" with our words made up on the strum.

His Excellency added his own verse:

I used to be a doctor and a jolly good doctor too,
But now I've finished doctoring I don't know what to do,
I'm growing old and feeble and I can doctor no more,
So I'm going to work my ticket if I can.
Chorus:
Back to Scott Base...happy land,
I'm going to work my ticket if I can.

The one thing I really wanted to capture was a decent group picture, but the mess room proved extremely tough to get a grand, all together shot. The dining tables were jammed together and the lousy flash unit did not have the buzz of a fly. The flash would not reach across the room to get a good

Scott Base leader Robin Foubister with Sir Arthur prepare to hoist the Vice Regal flag over the new Vanda Station.

one of the top table.

Next day, I had really great weather for the Vice Regal tour that would culminate in the opening of the new Vanda Station. Although Sir Arthur's predecessor Sir Bernard Fergusson, had visited Scott Base two years earlier, this time was special in that Sir Arthur was one of us, a New Zealander, Olympic Games medal winner, and the Queen's surgeon. We all enjoyed his company and genuine interest in what we were doing.

I was in one of two orange US Navy helicopters to leave McMurdo that morning for quick visits to the Scott hut at Cape Evans and the Shackleton Hut at Cape Royds. I was a busy boy snapping pictures at both places. Sir Arthur was very taken with the Adelie penguins at the Royds rookery, and I was able to get some good shots of him amongst the nesting birds before it was up, up and away to Lake Vanda.

I wondered if I would have enough colour film in the Canon. I had plenty of black and white film for the Rolleiflex but the 35mm was far better to handle and I juggled both colour and black and white. The trick was to note the number of frames shot, rewind, install the new film, do the same, put the original back in, wind on to where I was before and continue. The system worked, but I just had to pick the moment to switch. I needed the black and white for the media and the colour slide film for Antarctic Division library purposes.

After crossing McMurdo Sound, the helicopters ran into trouble heading over the Wilson Piedmont Glacier as the pilots were confronted with a wall of fog at the entrance to the Wright Valley. In spite of our valuable VIP

Historic moment after Sir Arthur officially opens New Zealand's first land based station in Antarctica, January 9, 1968.

cargo the pilots lifted their machines over the fog to 7000 feet above sea level to drop down into a rare and wonderful day further up the valley to Lake Vanda.

It was very crowded at Vanda, and I did not get any inside shots with the Governor-General. Those who could fit inside enjoyed a champagne toast with His Excellency. Without a flash unit I concentrated on the outside action which included raising the Vice Regal flag on arrival, the official opening, a big gang shot outside the front door, and, of course, the lowering of the Vice Regal flag on departure.

"That's the oddest place that standard has flown," I quoted Sir Arthur in the New Zealand media, as he described the Wright Valley being "out of this world and full of first class beauty."

In spite of the wonderful conditions (16 miles per hour wind and only -2° C) we had to be on our way as we had an at home night planned for the Vice Regal party at Scott Base. Sir Arthur enjoyed an extremely informal beer and chin wag with the boys in the sledge room, before he sent a message to the Queen from our

It was a nice day at Vanda for the Governor General and officials at the opening. An easterly wind at 16 miles and hour with a temperature close to zero Celsius.

Post Office, spoke on the radio telephone to New Zealand, mailed his philatelic mail featuring the prized Scott Base cachet and lined up with the rest of us in the chow line for the evening meal.

After that we all headed out to the Scott Base ski hill for an evening at the world's southernmost skifield. Previous parties had rigged up a rope tow to an old engine, and once the engineers had it going on all cylinders, we had this beautiful bowl to exercise our ski legs in. It was a superb evening and one in which we all relaxed as a "bunch of kiwi jokers" simply having fun. One of our Snow-Trac's had mechanical problems, so the young Porritts and Captain James enjoyed a brilliant six-kilometre ride home on the dog sled while Sir Arthur returned in the other Snow-Trac. The rest of us skied home. We got back late but had time to adjourn to the Mess Hut and catch the last part of the movie Those Magnificent Men in Their Flying Machines. It was a fun windup to a couple of extraordinary days.

The Vice Regal group's last night on the ice was spent at Scott Base. A bunch of us made sure they all got to their plane the following morning to give them a royal sendoff. Base Engineer Allan Guard was particularly thrilled that he got a half hour with Sir Arthur by driving him to the airfield in the British TransAntarctic Expedition's Sno-Cat Able.

Through Robin, we received a great letter from His Excellency the following week expressing his appreciation for all that we had done for him and:

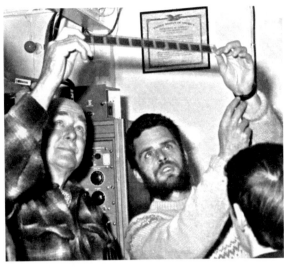

Sir Arthur reviews a photographic record with Peter Lennard, technician-in-charge, at the Scott Base laboratory.

This is just a line to thank you and your merry men at Scott Base for all you did for us during our all too short and vastly enjoyed visit. I am delighted we were able to have a night with you—although this is no way denies our great enjoyment of the excellent dinner the night before, even if we did have to return to the 'other place' to sleep. One has to live at Scott Base to really appreciate its wonderful spirit—where everybody will do anything for anybody else. And I am sure I don't have to tell you how popular New Zealanders are 'on the ice'—something that makes me feel very proud...

It was a delightful gesture of yours to send us on our way in the famous Sno-Cat (I will write in due course to Sir Vivian Fuchs, leader of the TransAntarctic Expedition in 1957 who drove the actual Sno-Cat) and Peter Scott (son of Captain Robert Falcon Scott).

January 18, 1969

Darling

The days have been very full. Thursday I waited up to meet some new arrivals off the plane. Got to bed around 3 am and was up again at 7 am to find that HMNZS Indeavour had berthed. I had time for a quick breakfast before Robin and I went to welcome the ship and meet with the Captain. I headed back to Scott Base with some passengers at 9:30 am, wrote up a couple of press reports about the ship arriving on its first voyage

The Governor General signs a guest book in Scott's Cape Evans hut.

of the season, and headed back to the ship at 11.30am to help with unloading, a time of all available hands to get the ship turned around and on its way back to New Zealand. We enjoyed lunch on the ship. Unloading finished about 4 pm. The work wasn't too bad really, but it was like something out of a Mack Sennett movie. We had the four Italians with us and somehow they got pie-eyed in the wardroom on rum and we had to remind them to get out of the way. I came back to Scott on the big truck and en route we had to change a tire. We had about 12 ton on board and it took us more than an hour to change the wheel. I did not get back to base until about 5:30pm. After a social hour and supper we headed out to unload the trailer containing all the frozen meat and vegetables and stack it in the snow cave, our natural freezer about a mile from the base huts. I went into the mine and stacked every bloody box as the others threw them down. We must have finished about 11 pm. We rode home on the sledge, singing our hearts out. We were very cold and very tired but warm within, if that makes sense. In the mess, we sang, drank tea, coffee or beer, made cheese sandwiches on the range and finally headed to bed about 1 am. At 7 am I was up again and stacked my newly arrived book supplies from the ship in the store. Six of us dressed up a bit for a 2:30 pm lunch invitation on the ship. When I got back to base about 4 pm I had about a hundred people wanting to buy books. I cleaned all that up and people congregated in my office and the Post Office for a happy hour. Guests arrived from the ship for dinner after which I had more book sales to the fellows coming in for telephone calls to New Zealand. After that, back to the mess to look after our guests. Well, a party developed and went off with

Sir Arthur checks in with a field party in the Boomerang Range.

a wild swing. Robin, Chippy, Taffy and myself were last to retire about 2 am a little the worse for wear!

This morning I had a wonderful sleep in, got up around mid morning and joined the party heading over the hill to church at the Chapel of the Snow. It was quite refreshing. Robin and I almost got the giggles as both of us were a bit husky in the throat and could not reach the notes. You know what hymns are like to sing at the best of times!

I bought myself a sweater from a navy officer. Strictly forbidden on his part of course. I paid him $8 for it. Not a bad price, methinks. It is a white polo neck Norwegian natural wool. However, I cannot wear it till I get home just to avoid any unnecessary questions.

Well sweet, as you can see it has been a full programme. I see a similar week ahead. My life on the ice is almost over and I am going to miss it all, and the blokes, but nowhere near as much as I miss you and the girls. Supper is over. Mmmm, not a bad supper tonight: stew, plastic spuds, frozen peas and beans, instant pudding and rhubarb. No one has yet matched the Sunday feed Wayne Maguiness and I cooked a few weeks ago. Each Sunday the house mice have to do the cooking to give Geoff a day off. I made some great hamburger patties.

Never been equalled as we were told it could not be done without eggs.

That's another thing! This week has been terrific. Fresh eggs! We have not seen those for sometime now. And yes, fresh potatoes. We all got pretty ecstatic over them. Funny where values lie, isn't it?

Happy polar chappie,

G

The ship's first supply run had the effect of reminding us summer-only fellows that our days in Antarctica were numbered. A favourite after dinner topic was around what we'd do when we got home. Some had jobs they could return to and then there was me. What was I going to do when I left here? I didn't have a clue what I would turn to though I knew there were some doors still open to me. My big question was what Lois and I wanted for the next part of our family journey. I really felt deep inside that Antarctica and isolation had changed my horizon. I vacillated between making a safe life in New Plymouth or finding something new. Besides, we now had two girls in school. Was it time to settle?

January 22, 1969

Darling heart

Rachel tries really hard, my love. Some days she is terrific and others she is most exasperating. Hilary gets her in their room and gives her printing lessons. She's very proud of that letter she wrote you. She had several tries as you could no doubt see. They both have turns helping me with the housework and are truly great to have. I really think it would have been worse if we'd had no children and you'd gone away.

This is how tomorrow's pay will go. The rates (taxes) will take $48, phone bill $8, food $10 (seems to last two weeks), maybe $9 to the dentist account. That makes $75 leaving $25 out of the pay of $100.30 (every two weeks). We are now only one payment behind on the house payments and the car is now down to a tidy $200 but I am not sure what the new charges will be. I'm trying hard to stretch the money round.

I thought that seeing as you have been away, instead of coming back here and falling in love with the place again I thought we could do all the adjusting together in another place. It will be odd for you to get used to this life again. I don't really know what would be best.

Lots of love from us,

Tuppy

January 26, 1969

Darling Husband:

I really want to come to Wellington when you get there. I could not bear to be up here with you spending a week down there before coming home. I'll just sit in a coffee shop and wait until you have finished your work.

You say that being down there is just like being in a dream. I'm sure it is the same with me at home here. Most of the things I do are mechanical and I do them 'cos I have to. Since you have been away I have found it easier to say 'no' to people. No silly nonsense, I'm just quite forthright. Sometimes this is a bit of a strain but my girls and I know what's good for us.

I have been persuaded to design the programme cover for Little Theatre's production of Thomas a Beckett.

Our veggie garden is yielding some nice peas, tomatoes and corn coming along. The only things that look a bit sick are some of the lettuces we planted.

Kisses from Us

Tuppy

January 25/27

Hello My Darling:

Well, t'is Sunday evening. We have just finished supper. Not much, just steak, peas and spuds. We sat around afterwards and had a quiet yarn while waiting for our movie to begin. Tonight it is Big Hand for a Little Lady starring Henry Fonda and Joanne Woodward. Should be good. Most of us are looking forward to it. It's been snowing all day and it is still snowing now. Beautiful, little soft flakes floating in the breeze – well hardly, the wind is blowing at around 30 miles and hour and the temperature about -10°C. Cloud covers the landscape and sky and land become one. Visibility is less than half a mile. So in some ways we have been shut-in all day.

I stayed in bed today reading till after noon! I washed my hair, had lunch, went down to Chippy's shop for a while and then wrote a wee story to go out on tomorrow's plane. It is a picture of the Mt Erebus crater for the Christchurch Star, taken by a US Navy photographer. I have asked the Star folk to send it out on the wire so you might see it somewhere. With our quiet day, Robin, Geoff, Noel and I spent the rest of the day till supper playing euchre.

John (Chippy) has just had a laugh at me as we wait for the movie. "Gawd, what do you write about." Now Hugh has just handed me a beer with the comment "You just sit down and write ... must be all bullshit." I must say I have been pretty spoiled the past few days. Thursday, Friday and Saturday I received letters from you. It was so lovely. I feel as though I am one-up on the others. Of course, I read some bits out to them. Your letters have been truly wonderful and always timely. I am top scorer for letters from my loved one. So I've received 42 and I still have them all.

Your explorer

G

January 27, 1969

Darling:

I took my art session to Pat's today and we had a great chat about travelling. She had a friend over from Sydney and we all came to the same conclusion that children are very adaptable. They might miss out on a permanent home perhaps but they gain a helluva lot. You know, when you are 40 it might be harder than ever to move on my love, so *I* think if we can move now we perhaps should. Don't despair about moving. Whatever we do will be terrific.

Your letter yesterday was full of high spirits. When you have a lot of work you always rise to the occasion and you are very happy. Right? Yes, you will miss the closeness of being with your South Pole buddies. What are you after when you come back? You are 28, and still supposed to be quick and alert in mind and body! Do you want to settle in New Plymouth for the next 10 years and then just vegetate after that? At 38! My love, *I* am not trying to talk you into anything, just trying to think of all the pros and cons. What type of work would you do at NZPA? Anything worthwhile or would it just be an ordinary reporter. Why not use your talents in the fields where you excel, industrial, oil and research? Yes, we do have a home here in New Plymouth that you more or less could say is ours, all the hard work is done anyway. Hilary most certainly would not go backwards if we moved. You know her mind is always leaping ahead. She sat up last night trying hard to read those two Antarctic books of yours. *I* am sure dear little Rachel would fit in anywhere, she has a complacent and cuddly nature. As for bruiser Bridget, she'd just win everybody's heart.

Much love and kisses

Tuppy

January 28, 1969

Darling:

Lately I've been nursemaid to Gin as she has been having blinding headaches. Not migraines but the doctor thinks it is something to do with her sinuses. So I am taking her off tomorrow to get some x-rays.

For a special treat I decided to take the children over to Mother's television to watch Disneyland. Howard and Gin and their kids came too, all in our car. Unfortunately I could not drive them home. As we were all piling into the car the dog got left out. I turned in the front seat to open the door for her, then wham, the door slammed shut on my fingers. Hilary said I was very brave as I only uttered "oh" instead of crying. Howard took me inside to the bathroom and I held my fingers under the cold tap. Then I got the shakes and went all white and queasy. It was the middle finger that really got squashed as it was the longest. The others just had a close shave! That mean't Howard had to drive our car. He managed very well with his knees up under his chin as I have the front seat right forward.

I'd still like to come and meet you in Wellington. I've become a bit possessive of you and I'd like to be there with you alone first. And then we can get back to our three girls, a basset hound and a cat!

We're getting excited

Tuppy

February 8, 1969

Hullo You Sweetheart:

I looked into a booking a motel for us in Wellington when you get there. Yikes, they are expensive! $63 for the week. I made the reservation from March 5 to March 12 so you better tell me if these dates are ok. Better tell me post haste so I can either cancel or not as our French lady is not running at present. I was on my way home from Mother's and I stopped at a Give Way sign when it happened. Everything graunched, ouch what a horrible word but that is what it sounded like. I managed to make it over the intersection and stop. Oil poured out from underneath the bonnet. It sounded like all the cogs were all out of timing, like what happened before. Let's hope it is not as expensive as the last repair job.

We all piled out and walked up to your parents' place. But they were out at a Bowl rehearsal so let myself in and called my Dad and he picked us up and took us home. Oh dear, what are we going to do. I called Neville (our Citroen mechanic) and he said he would pick the car up in the morning. Neville went ahead with the car repairs. It was the crown wheel and pinion and he also fixed the bushes on second gear. I asked him how much and wondered aloud how on earth we would pay for this. He said not to worry we'd have a chat about our sad car when you got back.

I had a long talk with Patrick on what you were going to do next. He reckons that six months back on the Herald and you'd be back in the old grind and getting nowhere fast and fed-up quick. You know my darling he's quite right too. What about thinking real hard on the matter. Wellington might be OK for a year with NZPA and then go overseas. And don't forget Canada!

Perhaps I had better not come to Wellington as it is going to be quite expensive.

Bridget's latest escapade.

We were staying with Esmae and Rob in Waverley. Esmae had visitors so lunch was delayed, also my checking up on Bridget whom I had tucked in for an afternoon nap. Wow, what a sight met me when I gently pushed the door open. One red, pink and black face child. My best lipstick, all dug out, my makeup squeezed out, the powder puff used to wipe it all up. What a mess! Then in my cleaning up, some of the makeup and lipstick mixture transferred to my clothes. Well, Bridget thought we both looked pretty. It took me a lot to control my mirth. There is such innocence in her beaming smile!

I am going to come to Wellington. To heck with the expense. Sorry you have had to read all about my ups and downs but you are more important to me than money.

Just before lunch today a realtor arrived at the door and brought a couple to look at our house. They looked it over and then left. I got on with lunch preparation when suddenly there was the realtor standing in the doorway. "They're most impressed and would like an option if possible as they are hoping to swap their mortgage but will not know until March 7."

My love, cross your fingers and send positive thoughts from your frozen home. Instead of a written option we just have a verbal one. We thought it best that way and he will let me know if they can't swap the finance. Price is $9000 or $9400 if they take the fridge, washing machine and lounge suite. It looks like the mortgage balance is $4950 plus about $1500 for second mortgage. So that will leave us with about $2500 to clear everything up and set us ahead for the next step.

Ally and Lance leave for Masterton next week. You will miss them. We had a great farewell for them. I am very sad at them leaving us behind though!

Four-in-waiting

Tuppy

Time was drawing near for me to leave the ice. With just a few days left, I still found time for a couple of extra-curricular outings. I was fortunate to get one last look at Lake Vanda and Vanda Station. I flew out with Robin for a quick one-hour stopover. I'd already sent a feature article and pictures to the New Zealand media and that meant this was a purely personal last look at the place that had been the object of a great adventure and to shoot some new pictures for the Ant-

My final look at New Zealand's Vanda Station, February 1969.

arctic Division files. Although there were one or two more chopper supply trips planned, I figured I was one of the final visitors until the spring resupply some six months away. The five men were now on their own.

I got to go seal hunting with Noel and Hugh. Under the Antarctic Treaty, New Zealand was permitted to gather 50 per year for dog meat. The guys brought in 16 yesterday. I could not really call this hunting and I saw first hand just how easy it was for commercial seal hunters. Hugh or Noel just walked up to within 10ft or so of a seal and shot it with a .303 rifle right into the heart. Death was instantaneous. One poor fellow in the colony died a mere 10 feet from another that made no attempt to move.

Nine of us took off over the hill for a final, nostalgic look at the past in Scott's Discovery Hut. We also had a great view of killer whales at play in the harbour. We headed over to the McMurdo goody shop (PX) to see what new stuff was available. I bought some gifts for Lois, the girls and family with money I had saved from book sales commissions.

I helped put together a summer support fancy dress party for the winter over guys. I dressed as a newspaperman with a newspaper shirt, hat and trousers. It was a happy sort of goodbye evening with Slim Dusty on the record player or our own memory lane session singing oldies of our youth, Tab Hunter, Elvis, Tommy Steele stuff. There was a lot of hilarity and who cared if I, or others, couldn't sing.

As our ship pulled away from the McMurdo dock
and out into the Ross Sea, the winter crew ran to
Hut Point for a final wave beneath Vince's Cross.

15.
ALL THE WAY HOME

I had trodden the icy wastes and spent five months as a photojournalist in New Zealand's corner of Antarctica. These adventures were heady enough for any young bloke. Include in that the thrills—and the fear—of visiting the South Pole, driving across the sea ice in the last of the big tractor trains, digging vehicles out of crevasses, using the purpose-built tractors which had completed the famous Sir Vivian Fuchs—Sir Edmund Hillary epic crossing of the Polar Plateau some 12 years before and you have tales to tell for the rest of your life.

I thought I had done it all until the offer came to shed my mukluks for deck friendly footwear for a voyage to New Zealand aboard HMNZS Endeavour. Wow, a chance to sail the notorious southern seas! This would surely be the grand finale. And I'd get to experience travelling across one of the stormiest oceans on the planet. I was looking forward to sailing home and shared my excitement with Lois on the telephone. "It'll be a grand trip," I said, "and a fitting finale to this fantastic adventure. We'll call at Campbell Island and then the Antipodes Islands."

I was scheduled to be the last of the summer crew to leave the ice on one of the season's last flights out to New Zealand. My plane seat was booked. But storeman Doug Spence needed a few more days to complete his work following the unloading of stores from Endeavour. Besides, he admitted, he was not a sailor and pleaded for the plane ride. The upshot was a switch I was very happy to accept. On top of that Robin designated me as leader of the sea-going summer group to liaise with the ship's captain Commander Doug

Summer and winter crews mingle on the deck of HMNZS Endeavour just before departure.

Bamfield and his officers. I was ecstatic! Once again, this was the stuff of my youthful dreams, and the nice thing was the voyage would not impact my original date of arrival back home as I would be leaving earlier. I called Lois and told her of the change in plans.

"I really want to come and meet you there," she wrote. "I can wait in the motel while you complete your work. I've really missed you and just want to be with you. I know it will be expensive but we'll find the money somehow. Besides you are more important to me than money."

And I wanted to be with her, too, to see her, hear her and touch her. I loved her plan but felt conflicted even as I told her I was excited. I was scheduled to spend a week debriefing before my time with the Antarctic Division ended, and between wanting to complete my time with the Antarctic Division on my own and having little in the way of firm job prospects, I was hesitant about time together in Wellington. I cheered myself up with the idea that with Lois in the capital, we would link up with old friends and check the place out in case a position with NZPA opened up.

Six days before the Endeavour sailed, though, I received a letter from Lois with the news that there was a possible buyer for our house. There would be a closing date of March 7, the same day as I might arrive in Wellington. And then, bam! I found a *New Zealand Herald* advertisement for a position as Chief Sub Editor of the *Fiji Times and Herald* in Suva.

The newspaper by then was 10 days old so I booked a radio-telephone call to Fiji and spoke with the general manager. I'm not sure I was believed at first, but after I insisted, the receptionist put me through to him because I was "calling from the South Pole". He took the call and told me how hot it was in Suva and added he was waving the phone around his office to cool the room down. We discussed the position and my interest and I ended the call

Out into the blue of the Ross Sea with Mt Erebus fading in the distance.

by adding a letter and clippings would be en route in the next mail out. His final comment was that while there had already been a lot of interest in the job my application would be on top of the pile.

I wrote out my letter that night, gathered together a few clips NZPA had sent me of my Scott Base work and put them in a very large envelope and patterned the front with the much sought after Ross Dependency stamps. Brian, Dave and Bob in the Post Office very carefully hand cancelled the stamps, added the Scott Base cachet to the envelope and popped it into the mailbag.

I definitely wanted my application on top. I phoned Lois as soon as I could on the schedule and shared the great news. She was all fired up about the possibility and was ready to do whatever was needed. Before leaving home I had assembled a package of clippings, the journalist's portfolio, and left them with her for just such an occasion. The next day she had these into an envelope and off to Fiji to backup what I had sent through. Now, we would wait. Because I'd be at sea, she was the contact.

The possibility of a two-year contract in the tropics helped modulate my dark thoughts about being homeless and out of work. It was ironic that I might move from the South Polar climes to the equator! And it was pretty darned exciting.

Relaxing in warm sunshine on the after deck, from left, Bob Hancock, Bruce Dowie (en route from Campbell Island to Lyttelton), Hugh Clarke, Alister Ayres and John Newman.

Meanwhile, all hands were at work on clearing the ship of the last of the supplies. I also had some last minute public relations activities to wrap up including the first-ever international Antarctic bicycle relay race from the ship to Scott Base. Three teams competed: USARP, NZARP and Endeavour. The ship provided the bikes. The Scott Base crew won the event with the last rider carrying his bike over the finish line after the front wheel fell off.

On my final night at Scott Base, I scheduled a three-minute call to Lois within a crowded lineup on the radio-telephone link. It was a quiet call, one of those times when we just wanted to feel each other's presence. I promised that if it was possible I'd call from the ship, or at worst, as soon as I got to Lyttelton. With just hours before departure, I had to finish up a couple of stories about the close of summer activity and get pictures of the winter-over crews at Scott and Vanda to the Antarctic Division.

It was time to go. My gear was packed and thrown into the Landrover. Farewells would happen at the ship

- Darkroom clean and tidy for the winter guys? Check.
- Book sales inventory cleared and handed over? Check.
- What about my desk? Did that look OK? Check.
- My room? Everything picked up, cleaned? Check.
- Me? I'll stay dirty. There's a shower on board.

For the last time, I drove the four miles to the dock amazed and jubilant that I would sail home with the Royal New Zealand Navy; in a bit of a daze about a possible job in the Fiji Islands.

Robin and I joined Commander Bamfield for lunch and although I had the option of accommodation in the officers' quarters, I chose to sleep and eat with my colleagues up the front end. There was one cabin there with bunks and I shared that with Chippy while our mates got to sleep navy-style in hammocks. I stowed my gear, and went up on deck while the ship made final preparations for sailing. The winter-over guys mingled with us until the last whistle. In spite of the very manly, handshaking farewells, I could feel the emotion. The gangplank was hauled aboard, the ship cast off and slowly she edged away from the wharf into open water.

At 1:45 pm on February 22, 1969, the ship cast off and the winter boys scattered from the dock. It was not long before we knew why. They hiked up nearby 700-ft Observation Hill and waved to the ship as she rounded Hut Point and moved through the brash ice to the open water of McMurdo Sound. I felt tears at seeing these figures standing, cheering underneath the Vince Memorial Cross. I choked as I recorded the scene on film as we waved and watched this remarkable part of Planet Earth escape into the distance, knowing those wintering over would not see another supply ship for almost a year. And, when would we ever see them again? Robin, Allan, Peter, Noel, Wayne, Geoff, Keith, Nigel, Doug, Brian, Dave, Chris and the Vanda team, Bill, Simon, Al, Tony and Ron. My shipmates and I were all very quiet and withdrawn. I couldn't believe it. This was a moment I had not expected. I'd shared my life, dreams and aspirations with these guys. Our lives had depended on each other. And yet, with a wave of the hand it had ended.

The shoreline dropped slowly away under the grey skies. It was like a big black curtain coming down on this stage. And there'd be no encores. The separation was brutal. At least, when I left Lois and the girls we'd still have contact. Above all, I knew I'd be back to laugh and play and grow with them.

We turned our faces northward as the islands in the bay slipped by. With Hut Point diminishing in the background, I picked out the final landmarks we'd grown so used to: Mt Erebus, Mt Bird, Mt Discovery, White Island, Black Island, the Barne Glacier. A knot in my chest, I stayed alone on the after deck for some time listening to the sea swishing by. I was close to tears and wondered what this place had done to me. I projected ahead to the reunion that was in store for me with Lois and the girls. What effect had my absence had on Lois and our daughters? Our relationship? My pay cheques would end very soon. What would I do now to put bread on the table? I wanted to be home, but I did not want to leave this place and time I had come to know and love. Would I be able to once again fit in? Lois had already mentioned this. Sitting at the stern of a naval vessel slipping through calm water did not produce any answers.

With me on the ship were Alister (Taffy) Ayres, John (Chippy) Newman, Bob Hancock, Derek Cordes, Hugh Clarke, Bruce Brookes and Charlie Hughes. I found them all below decks in their new quarters rejoicing and celebrating being clean. I took their directions to a blessed, long, hot shower followed by the comfort of fresh, clean clothes. O, what luxury!

With wonderful and pleasant sailing through the indescribably brilliant blue waters of the Ross Sea, we easily slipped into a sort of happy, dreamy state, nothing to do and all day to do it in. Quite a contrast to the pace at Scott Base, field trips, visitors and our work, the reason for being there.

I awoke most of the days on board to the sound of the ship's pipes then enjoyed a beaut shower, hot water and plenty of it. I hadn't realized water could be such a treat. The ship continued north, and during the first days we gently rolled, rose and dipped in the unreal salty blue, cameras out photographing the ginormous icebergs.

Far from being a sleek frigate, HMNZS Endeavour, was purely and simply a workhorse in her role was as an Antarctic supply ship, delivering fuel to research bases in Antarctica. She carried more than a million gallons a year to McMurdo Sound alone. The ship was built in 1944 and commissioned as USS Namakagon, to transport gasoline to US fleet warships in the Pacific in the latter days of World War 11. After that she served on the west coast of North America before being mothballed in 1957. In 1962 she was transferred to the Royal New Zealand Navy and commissioned as HMNZS Endeavour, the second RNZN ship to carry the name of Captain Cook's ship. HMNZS Endeavour was returned to US custody in 1971.

The further the ship sailed towards the South Pacific ocean, the more feathered friends delighted us with their dipping and diving above the wake and gentle swell. Our gathering place on the after deck gave us a great view of the snow petrels and Antarctic petrels and in the early evening of the second day we saw our first Sooty Albatross. I shot plenty of film hoping for at least a couple of magic photos.

The First Lieutenant told us that the birds we'd identified as Antarctic petrels were in fact Cape pigeons. We took his word for it as he had done this trip a few times before. About 16 inches long, this pigeon breeds on most of the sub-Antarctic islands south of New Zealand and spends most of its life at sea. Watching these black and white speckled birds sail just above the sea became an enjoyable pastime as I idled away the luxury of time without deadlines.

The bird that really captured my attention though was the Sooty Albatross. A couple of them continued to soar majestically around the ship and over the wake. They just floated on the wind, feathered gliders. There was no land in sight. We were like them, a speck on the rolling ocean. It was a rare treat to just stand on the quarterdeck and watch an albatross hoist himself into the air. These giant birds would run along the water, up a wave and (hopefully) take off as the wave crested. It was quite the technique but as I noted, not always successful. The light mantled sooty is small by albatross standards with a 7.5 foot wingspan and 35 inch body. In this area of the southern ocean these winged beauties nest on Campbell, Antipodes and Auckland Islands.

Now, a new variety of albatross joined us, white with a yellow beak, probably a southern royal that breeds on Campbell Island. The southern royal albatross though is a giant with an almost 10 foot wingspan supporting a 48 inch body. The royal's largest breeding ground is Campbell Island. These birds spend more than half their life at sea, only going ashore to breed.

Our fourth day out, the 25th of February, I followed my daybreak ritual then piled up on deck and sprinted aft to our day quarters and after breakfast stayed on deck gazing at our rolling saltwater surroundings. We'd seen the last of the icebergs. There was a perceptible change in the air. The forecast was for gales. My all too brief log notes read that I was "looking forward to this".

We had a great top to bottom tour of the ship including a visit to the engine room. In the afternoon, we watched a Beatles movie that our resourceful and shipwise Bob Hancock had organized. Outside we experienced rain...

Wandering albatross soar and glide above the southern ocean. These birds followed the ship for several days.

glorious real rain...the first we had tasted in five months.

Our fifth day and the sea started to get moody. The ship had a bit more roll and more rise and fall in the heavier waves. The storm forecast was right. I spent the evening up on the bridge. The barometer continued to fall. Visibility was down to about four miles. The ship was headed into a gale as it entered the whaling zone of the southern ocean. It was raining and we had a wind of about 20 miles per hour on the port bow. While I was pretty excited about the coming storm, I was so-o-o-o glad I had the comfort of being with the Navy! I stayed on the bridge with the Second Lieutenant till about 9.30 p.m. watching the bow rise and crash into the waves, spray spumed back across the foredeck. I went to share this excitement with my colleagues and found them playing Scrabble and Monopoly. No games for me tonight though so I headed to my bunk, quietly thrilled to be sailing through a southern storm.

A massive thump and crash woke me around 3 am. The ship was rolling and pitching heavily. I wondered what we had hit...maybe we'd gone under, or hit an iceberg! Naaah. I called to John but he slept on. I swung my legs over the side of the bunk and dropped to the floor that seemed to disappear from under me as the ship rolled. I hit the floor hard and holding on to the cabin wall dressed to go up on deck and to the bridge as I did not want to miss any seaboard action. Fully awake, I realised we were still afloat and, being in the bow, the crash and thump was amplified as the ship ploughed bow first into another wave. Outside the cabin, the hammock boys swung with the ship's motion. No-one answered my whispered call, so I went and enjoyed my shower routine in spite of the rolling and pitching. I escaped to the bridge for the best view. I was completely awed by the raging, tormented sea. The officer on watch reckoned the seas were only slight. The wind had increased to 25 miles an hour. As we nosed into each wave, the spray was flung back to crash on the bridge windows. Nervous excitement tingled in every fibre of my body. I was thankful to be travelling in a ship that had made this voyage many times.

Breakfast called. But to get from the front end to the back end, I had to run across a catwalk over the deck. With every roll of the ship, the waves crashed across the catwalk, obliterating the stern. The only way across was

Smashing seas swamp across the Endeavour and she rolls
through a wild south Pacific Ocean.

to gauge the timing between each roll and wave and run like mad before the next wave crested over. I made it that time. But if this kept up I could see we'd be spending a lot of time in our mess. Breakfast was a true experience this day, not because of what we ate but how we had to eat it, given the turmoil of the ocean. This morning it was mince (sloppy joes) and currant buns. It was tough keeping the food on the plate though. Some of the guys used dessert bowls to hold their meal in one place. We had to hold tightly to the plate. John, normally a big eater, could only manage a couple of the small and very tasty currant buns. Like several of the other guys he was feeling a tad seedy. As a diversion Bob went off and organized a movie, Warning Shot, with David Janssen.

The waves were way above us now. In my mind I was trying to figure out how to best capture this on film. Staying up on deck was good, just being in the wild storm turmoil and fresh air reduced any nausea. I tried to encourage a couple of hammock-bound colleagues into coming on deck and sampling the brisk, fresh sea air but they would not have it and told me just where I could go! Lunch was a real effort, but the captain very thoughtfully altered course for a time to put the ship into the wind to steady her and provide a degree of comfort so that all on board could get a good bite to eat. That afternoon we tried to play Monopoly but that was a lost cause as the bits flew everywhere. Out on deck with the wind and the waves, I did not see any of our sooty companions but the royals stayed loyal, a lot of petrels as well.

The ship was now rolling through a good 20 degrees. I went to my bunk that night supercharged after a fantastic stormy day. It was impossible to hold a book steady and read, so I wedged myself into the bunk between the hull and the side rail, and flipped out the lights early at 10pm and fell asleep. In spite of the howling storm outside I slept well, and at daylight found our ship fully in the play of the gale. Back home Lois and I, Hilary, Rachel and Bridget, would often drive down to the port when the weather turned rotten so that we (read me) could enjoy the crashing waves upon the rocks and watch the seas surge and smash over the breakwater. As young teenager, I'd often found myself in trouble with the port authorities, who'd find my friends and I attempting to run or cycle along the three-quarters of a mile length of the breakwater dodging the spray and the waves as they crashed over. It was a totally stupid thing to do to slip through the safety barriers, but I loved the rush of the challenge. The real danger was being washed off the breakwater and into the harbour. Now I was right in it, at the mercy of wind and wave but in the comfort and safety of the Royal New Zealand Navy. What more could I ask?

Commander Bamfield invited me up to his quarters to inquire about the well-being of the team. He shared that this was the worst storm he had encountered in his eight years of taking the Endeavour on the Scott Base supply runs. Wow, I thought, this was a story to share with Lois and the girls! The photographs I was taking now would be for them. It was even more challenging this day to get from the bow, across the catwalk to our day room and food at the stern. Amidships was almost permanently awash in the pitch and roll. Estimating the sprint was the game of the day. We had to go one at a time and this gave the folks at either end a jolly good laugh at my dash when I miscalculated the moment and got absolutely drenched.

I spent more time up on the bridge watching and listening to the violence of the sea. At one point, the ship was steaming at full ahead yet barely had sufficient forward power for steerage. We were punching into an ocean current that was all but pushing us backwards. As I came down a gangway towards our quarters to check on the hammock-bound boys, I lost footing and smashed my head on a steel bulkhead. I dropped to the deck with blood everywhere. A couple of the guys hoisted a groggy me from a crumpled heap on the deck and hauled me off to the ship's doc who decided the wound did not need any stitches. He gave me a couple of painkillers and told me to lie low, relax and take it easy.

Although sleep had been tough for a couple of nights now from the ship's rolling and pitching, I was suffering re-entry nerves as the return to normal life inched closer with each passing day. Some of the guys had jobs to go to and some did not. I called Lois a couple of times on the ship's phone to see if there had been any developments in our hope and desire to land a newspaper position in the Fiji Islands. At the beginning of my polar adventure NZPA had made an informal offer and maybe this could be possible employment for me. Before my trip south the previous October, the managing editor confirmed our discussion "to make some recognition of your work on the basis discussed." I remained not quite sure about wanting that position though, as it would mean moving to Wellington.

All night we felt the bash of the bow into the waves and the spray raining on the deck. This was life at the sharp end. I did not expect much change in conditions until we reached the waters off New Zealand.

I thought about this voyage and how we had calmly sailed out of the brash ice into clear blue seas before being thrust into the teeth of a wicked storm. Is this the metaphor? What do I face when I get home? Lois and I had agreed to this adventure on a financial shoestring. At home, I doubt if there was any string left.

Good grief, I thought, if the *Fiji Times* position eventuated we'd have to sell the house, pack up our stuff and move with three children to a tropical island. I recalled how Lois and I had lain in the tall grass on the farm watching the clouds go by and thinking of what we'd like to do when we were married. Life in Fiji had occupied many a romantic dream. Now there was a chance it might happen. Lois was very keen and encouraging, even though it would mean a big upheaval. She told me on one of our calls that the newspaper owner wanted to interview her and would come over from Australia to visit her in New Plymouth. I learned that the owner's rationale was that he had to be satisfied the wife was cut out for life in the tropics and would be able to cope with the cultural differences. But there was no indication of where or how I would be interviewed. Being at sea, and still many days away from the front line action on our future, I was doubly troubled with events happening without my participation. A truly odd feeling.

I was deep in my misery on our deck at the stern watching the waves and the ship's wake disappear into the clouds when I heard my name over the public address to report to the radio room. I had a call from Australia. The Fiji Times owner was calling, and over the crackly ship's telephone, asked me a couple of questions and then declared I had a job! Misery departed with the next big wave to smash over the bow. I had some difficulty hearing the full details over the radio-telephone but what the heck...I had a job to go to.

Elation surged as Endeavour shook her way up the next wave and doubt again returned as she crashed into the trough. Is this right for Lois and the girls? Sure, it had been a dream of ours for a long time but this was different. It was happening, leaving home and family for something quite unknown. I called Lois and together we teared up at the jubilant news. Now, a new adventure for all of us.

This signpost with a smile points straight down Perseverance Harbour on Campbell Island. Tierra Del Fuego is the next land point, 5328 miles across the Pacific Ocean.

Next day the seas showed some moderation under heavy grey skies that slowly gave way to a glorious blue. The wind was still fresh at about 20 miles and hour but for now we had an enjoyable break from the tormented rhythm of the past few days.

Campbell Island appeared on March 1 in a cold, grey dawn. Our visit here was to drop supplies and personnel and pickup a trio returning to New Zealand.

This was another place in my private adventure gallery that I now had the opportunity to at least visit. I got up early at 5 am to clean our galley and hopefully get a sunrise picture as we approached. Not to be. Big seas crashed all around us. And rain. About 7 am we headed into the island's Perseverance Harbour. The weather was so foul I figured the best pictures might come later in the day on the outward journey. Just inside the heads, we were hit by willy-willys, wind roaring down from the hills with gusts around 100 miles and hour. We could watch them flatten down the hillsides and swirl across the harbour before feeling their impact as they slammed into our sturdy little ship.

We anchored off the New Zealand weather station in Beeman Cove and I was fortunate enough to be in the first boat to shore. I met the blokes on the station and as this would be an all too brief visit, Chippy and I went for a quick walk. Man, did it feel good to be walking on dry land though we had those wobbly sea legs and marvelled that the ground did not move beneath us. We wandered through towering scrubby growth to get a glimpse of the harbour and anything else we might see:

- Flowers in bloom in this sub-Antarctic environment.

- The joy of seeing green again

- A sea lion in the undergrowth

- The rocky ramparts in the centre of the island

- A wonderful view of the shoreline

- Views straight down Perseverance Harbour under the signpost pointing the way to the next landfall east, Tierra Del Fuego, 5328 miles across the Pacific to South America.

It was a hurried walk through the arched undergrowth but very satisfying with little time for photographs. All too soon we had to hurry back to the ship as we could see signs of departure at the wharf and did not want to be left behind. We reached the boat in time and powered back to Endeavour. The weather had deteriorated further and so I shot pictures for mood as the ship set its course for a similarly brief stop at the Antipodes Islands, about almost 500 miles slightly north and east of Campbell and about the same distance southeast of New Zealand.

Like Campbell, Antipodes is the remnant of an old volcano and provides an ideal breeding ground for the sooty albatross and cape pigeons. The plan was for Endeavour to call at this wild-windswept, exposed rock to pick up a research party but it was just too rough and tortuous for a landing party. We did stooge around for a bit looking for a break but eventually we had to move on leaving the pickup for a later ship to call. I wondered how the fellows on the island would feel about that. Between a foaming sea and low cloud, the conditions were so bad I could barely make out the island from the bridge.

After leaving the Antipodes, we headed straight for Lyttelton, New Zealand. In a couple of days, we'd be home and a very different future. Soon we would all scatter, five to their homes in the South Island and three of us to our homes in the North Island. New topics found a way into our discussions, especially "I wonder if we will see each other again." We created a small diversion in creating an Antipodes Star for Jim Steil, a young US Navy fellow who had joined us for the voyage home. We presented it to him as we entered Lyttelton Harbour, conferring on him honorary Kiwi status in celebration of his maiden sea voyage even though he had been in the US Navy for a few years.

We'd packed our bags, and made ourselves presentable for reintroduction to New Zealand society. I looked at each of my fellow OAEs as we gathered on deck. We were dressed and pressed, sporting ties, dress pants and our monogrammed black Antarctic jackets. With them, I had added my voice to the Antarctic tableau. My footprints were there and would remain. My summer was now frozen in history. A piece of me would always be there. Together with the fellows who were wintering over, we had opened New Zealand's second decade of Antarctic activity. We'd used our talents, ingenuity and resourcefulness to establish our country's first mainland winter over base at Lake Vanda. We'd added our support to a variety of ongoing university programmes, we'd hosted international groups and consolidated New Zealand's role on the planet's last wild frontier.

Lois too had faced into an adventure caring for our daughters, our home, learning to drive, branching into teaching a group the art of colour and texture in painting and taking her own talents into theatre.

Together, we'd head into the next adventure with courage and confidence on a tropical island. And after that there'd be another . . .

Hills and houses of home. HMNZS Endeavour sails into Lyttelton Harbour. From left: Derek Cordes, Bruce Dowie, Alister Ayres, Hugh Clarke, Jim Steil (US Navy), Bob Hancock, John Newman and Charlie Hughes.

All set for a new life in Suva, Fiji Islands, March 1969: Hilary, Lois, Bridget and Rachel take a look at a Fijian bure (traditional house).

EPILOGUE

Our journey has been remarkable. We fell in love with Fiji the second our DC3 tyres scorched on to the runway at Suva in bright and hot mid-afternoon sunshine. The newspaper editor, Len Usher, and Suva's mayor and his wife greeted us and whisked us to their house for a meal where we relaxed. In 15 days, we'd completed the sale of the house, organized a household move, organised all the necessary documents, completed the farewell rounds of friends and family, and headed out on the first stage of the trip from New Plymouth to Auckland where we stayed overnight with Alister "Taffy" Ayres and his wife Merle, whom we met for the first time.

Fiji was the perfect place to grow into the new life we had set before us. My job was all and more than I could have expected to follow a correspondent's life in Antarctica. Within just a few weeks of settling in, I experienced what it was like to truly be a newspaperman. In July, we found the Fiji people were totally into the Apollo space programme. And that prompted me to be at the paper well past midnight transcribing news off the shortwave radio, writing a story and producing that day's headline of the successful lunar landing by Neil Armstrong on July 20. There were people lined up at the door at first light to get newspapers. And I got what photographs I could from global news services to supplement the wire copy. A few months after that, and we had a Royal Visit of Queen Elizabeth and Prince Phillip, complete with HMS Britannia at anchor in Suva Harbour. Our *Fiji Times* excelled in the coverage, complete with a special edition. As a family, free time became beach time and investigating life in the tidal pools. We were absorbed into the crossroads of Fijian, East Indian, Chinese and western cultures. Sadly, our idyllic blue ocean and coral reef adventure came to a close with me falling to severe dengue fever. But that prompted us to set our sights on emigrating to Canada after we completed the two-year contract in Fiji.

Once in Canada, our family adventures included camping trips throughout Alberta, Yukon and Northwest Territories and British Columbia. We raced a dog sled in the Yukon; I cross-country skied from Saskatchewan to Alberta and cycled the Rockies and Nova Scotia. Although we did spend part of the 80s in New Zealand where I left newspapering and began a new career in oil industry public relations, the oil business returned us to Canada, to exciting new projects on Canada's east coast and then to corporate headquarters in Washington DC in an international public relations role. When all that was done, we again returned to Canada.

Lois and I have climbed the mountains, walked the valleys, skated on thin ice, and with the help of family and friends hauled ourselves out of crevasses. In spite of the rocks and boulders, we have remained strong. In April 2011, we celebrated our 50th wedding anniversary, a brilliant family event at a western ranch resort in the shadow of the Canadian Rockies, just east of Banff. Technology allowed us to share a part of the event, the renewing of our vows, with friends and family in New Zealand and several of my Scott Base mates, all of whom surrounded Lois and me in a time I see now as our making.

In April 2011, we wrote this story about our latest adventure and sent it out to friends and family:

Burgess Park (New Plymouth) to
Fish Creek Provincial Park (Calgary)

Day One
She offered the thought to visit Heritage Park and dress up in old-fashioned costumes for a surprise picture for The Event. Neat idea, said he. Let's do it.

Day Two
They drove right past Heritage Park. You see, they forgot.

Day Three
"Why don't we go in," said he, at the traffic

lights. It was snowing. It was windy. Chilly.

"Ok," said she.

They wound round the Heritage Park roads,
angling into a snowbank close to the building.

There were few people about. He and she hustled
over to the Pioneer Photo Shop. Closed.

"What time does it open," they inquired at the store next door.

"Only on Saturdays, " the young pioneer-dressed lady
responded. "Weekdays by appointment only."

"We only wanted to find out what they offer," he mumbled.

"Wonder how much they'd charge,"
said she getting in to the car.

Day Four

"I think that would be too expensive," he said
next day. "Maybe we could just dressup a bit
ourselves and take our own picture."

She leapt at the idea. It took shape. It had form.

Check the camera, tripod and flash. Mock setups
in his office to ensure self-timer OK.

Day Five

They went shopping. Value Village. She bought an
old lacy white tablecloth. "Beautiful," she said. $9

Next stop Fabricland. Some white stuff to put
under it and prevent show through. $3.

Home again. She is on her knees making a pattern,
figuring out how to cut and match bits of tablecloth
for a snappy white dress, fluted and flowing.

Day Six

The sewing machine whirred. Semi-naked she
strode busily from workroom to the full-length
mirrors. Pinning, matching, fitting. Success. She
had a dress. She looked radiant, beautiful.

He tried a couple of old suits. Suck the
tummy in, ahh the pants still fit.

Dollar Store next for fake white flowers. $1. Back home,
final stitches, a hippie flower halo is shaped. Terrific.

Day Seven

Sunny afternoon after church to Fish Creek Provincial Park to scope out likely sites for their photo. They walk their traditional pathways through the natural forest, stop, pose and photograph. The pictures are reviewed on the computer. Four sites selected. Check the weather forecast. Fresh snow promised.

Day Ten

Still snowing. By noon it stopped. Still overcast. Nice soft light. Dressed for a wedding, she clutches her magnificent white wedding dress above the snow. They're wearing winter coats, mitts and boots.

Location #1: In deep snow around a fallen aspen tree. Looks good. Bit of trouble positioning on the tree.

Location #2: More formal under a spruce bow. They stood as they had done before. Flash worked best.

Location #3: Seated on a fallen spruce tree. She drapes her dress just so. He scrambles from camera to tree, sit one, stand one, leap over the tree (fall over) and stand behind. They're laughing and having a wonderful time. He looks through the viewfinder. Man, she looks gorgeous.

Location #4: They drape blankets over a park bench. She changes her shoes one more time and poses on the arm with her feet on the seat. "Just great," says the viewfinder. He presses the shutter and dashes to pose.

They pack up the gear, dress warmly again and, holding hands, head to the car. In the warmth, a quick review shows shoots 3 and 4 to be the best. Excited, she leans over and kisses him.

April 8, 2011

With a golden anniversary in the bag, Lois proclaimed
our next target. "Just 10 more years," she laughed, "and we'll
be 60 years and you can buy me a new diamond."

After almost five months of perpetual daylight, it was terrific to record a real sunset as HMNZS Endeavour made its way north from McMurdo Sound towards New Zealand.

CRYSTALS

It is not the mountain we conquer, but ourselves.
--Sir Edmund Hillary, conqueror of Mt. Everest.

It was only a few years ago that my youngest granddaughter Gillian proudly advised her teacher during a discussion on penguins, that "my Poppa has been to the Antarctic" and suggested I talk to her Grade 2/3 split class. In her seven-year-old mind she considered she had the teacher's concurrence and rushed home to make the invitation to come to her class the following week. I went ahead, dusted off my little rock box and gathered a few slides of Adelie penguins, seals and ice. Gillian was very excited but when she checked with the teacher the day before I was due it appeared the teacher did not believe that her grandfather had been to Antarctica and been face to face with penguins. She came home crying. My daughter jumped into the fray, contacted the teacher and vouched for her daughter and her Dad. With apologies, I was scheduled in for the following week. I arrived at the school with projector and laptop for backup. Gillian was absolutely thrilled when she was called to the office to meet me and, hand-in-hand, proudly take me to her classroom where I was given a rousing welcome.

The youngsters were thrilled to see my slides of the penguins I'd seen 40 years earlier. Time did not matter. They asked questions and stroked and touched the specimens, especially the penguin wing. The one-hour presentation turned into almost the afternoon. It took me back a long way since I had not had a classroom discussion since our own girls were in elementary (primary) school. The reactions that afternoon served as a kicker to get serious about recording in words and pictures my brief encounter with the south polar icescape. This

271

book has found life because of the interest, kindness, help, advice, criticism, encouragement and support of many people. I've been surrounded by family, friends and acquaintances who wanted to know more, not only about Antarctic exploration but also its impact on a young family from post-war New Zealand. Apart from making an occasional reference to the five-month Antarctic sojourn, I'd relegated the adventure to simply one of those life events in the "been there, done that" drawer. When my eldest granddaughter Veronica, now at university, kept badgering me and asking questions about my Antarctic adventure, I realized that maybe I should give it a try. And always at the front of the line were Lois and our daughters Hilary, Rachel and Bridget who had never given up hope that one day, yes, perhaps one day, my stories would go beyond those snorer family slide nights.

My first thoughts were for a simple book of pictures and captions as a memory for the grandchildren. However, spurred on by our first ever, and probably only, New Zealand Antarctic Research Programme (NZARP) 68/69 reunion in October 2008, I ratted around in my boxes of stuff and uncovered a heap of material including notebooks, diary notes, news clippings, letters and photographs. I saw a larger story for the kids. My memories were as vivid as yesterday, and the book took on a life of it's own.

With two "chapters" drafted, I registered for a Writing It Real conference aboard an Alaskan cruise ship in early 2009. This had a dual purpose of allowing me to take Lois on a sea-going adventure while I, for the first time in my life, associated with other creative writers. This, I reasoned, would tell me if I had the right stuff. Although I'd spent my whole working life writing words for others, I wondered if I had a voice of my own. The encouragement from faculty and writers was inspiring.

When my own memory and memorabilia stumbled, I called on many of my old team mates to help fill in the gaps and check out facts. It has been enjoyable to send out the emails, and energizing when the replies came back at lightning speed. From New Zealand, Allan Guard, Christchurch, Robin Foubister, Muriwai, and Hugh Clarke, Alexandria, proved my mainstays, and augmenting their knowledge, diaries and memories, I've been in contact with Bob Hancock, Takaka, Keith Mandeno, Auckland, Brian Hool, Auckland,

Noel Wilson, Wanaka, and Chris Rickards, New Plymouth. I have valued the contact and input of Alister Ayres of George Town, Cayman Islands, Simon Cutfield, Brisbane, Australia, and Ian Stirling, Edmonton, Alberta, Canada. I felt intensely the depth of our friendships stemming from a few weeks together at the bottom of the world.

Marcello Manzoni, Udine, Italy, was with us as a geologist back then, and I have valued his input and interest in these past couple of years. All in all, Marcello has shown just how international Antarctica is. Since our 2008 reunion, we've enjoyed more exchanges and pictures than we ever did on the ice.

I was thrilled the day Sheila Bender of Port Townsend, WA, founder and publisher of Writing It Real (www.writingitreal.com), agreed to edit my manuscript. Her craftsmanship and experience as a writing teacher, writer and author has brought the book a long way from my early drafts. Her coaching and suggestions have rounded out a story I threw in the too hard basket on several occasions. Her substantive developmental edits, notes and telephone calls put the book between its covers.

Kim Staflund and her team at Polished Publishing Group have been just terrific to work with to bring the manuscript to life. Kim's ever-smile and encouragement meant that publication was never in doubt. Thanks Kim. To John Rempel, my thanks for his exceptional work and patience on the cover and interior design, integrating visual interest to this personal story.

Sir Charles "Silas" Wright was one Canadian I did not get to meet. Silas was a member of Scott's polar expedition in 1911, but by the time I contacted his family in the mid-70s, this famous Canadian had died. A couple of years ago, I made contact with his grandson, Adrian Raeside, a New Zealand-born Canadian cartoonist, creator of The Other Coast. Adrian had just published Return to Antarctica, based on Sir Charles' diaries, a fascinating alternative read to the many books published about Scott's famous journey to the South Pole. Email has provided a great bridge of encouragement for our respective endeavours.

The friendship my wife Lois and I forged with Gin and Howard Marriott and Ally and Lance Girling Butcher continues to this day. Our gatherings are always filled with good laughter about the lives we've led and the adventures each new day brings.

There are those I appreciate for just being there and taking an interest in what I've been up to, knowing that a book is in the works. My thanks to Margo Moore, Eric Davidson, Terry McKinney, Carrie Sousa, Lesa MacPherson, Renato Guevara, Rod Matchett, and Rob Burns.

Grandchildren Gillian, Emma and Beth wander into my office at times and drape their arms on my shoulder as I tap away at the computer. "What are ya doin?" or "How's it going?" Robbie, Veronica, Ethel and Fisher inquire and encourage, though I'm sure they've wondered if the book would ever make it. Max, the eldest of them all, is already set in his adventure of big boat sailing in Northern Australia. I wonder where the adventures of the younger grandchildren lie.

The gift of family is a treasure and I'm glad to have shared many of life's moments with brothers Michael, Patrick and Brett and to be dazzled by my little sister Fiona whose big adventure took her to university in her middle years, gaining her Masters degree in Fine Arts.

I don't think I can really say thank you enough to Lois' twin brother Lynn, and his wife Robin, or Lois' sister, Esmae, for being there for us when we really needed it. Together with their families they have been a heartwarming part of our journey.

My dad taught me to how to make something out of what I had; my mother taught me the magic of words and the wonder of reading. Lois' dad showed me that garden dirt produces wonderful things, and her mother showed me the value of tolerance, patience and acceptance.

My love and my thanks to two others, my late elder sister Marie and my late brother-in-law Rob: I promised them I'd write a book. To them I now say "it is done."

And finally, in 1976, Lois and I set our feet on the Christian pathway. Glen and Marlene Paulsen of Grande Prairie, Alberta, provided the friendship and early knowledge of Jesus Christ that makes us who we are today.

42 YEARS ON

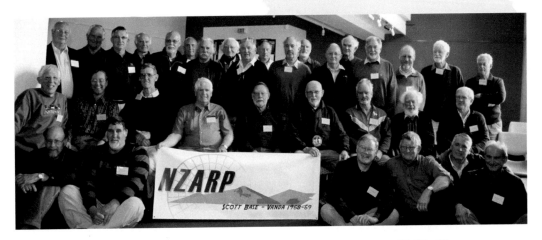

New Zealand Antarctic Research Programme 1968-69
40th anniversary reunion, Christchurch, October 17-19, 2008.
Back row from left: Miles O'Malley, John Whitehead, John Gemming, Charlie Hughes, Brian Hool, Alister Ayres, Ron Craig, Tony Taylor, John Newman, Hugh Clarke, Barry Kohn, Don Robertson, Dave Blackbourn, Euan Young, Bob Hancock, Trevor Crosby, Geoff Tunnicliffe, Wayne Maguiness.
Centre Row: Graeme Connell, Al Riordan, Simon Cutfield, Allan Guard, Bill Lucy, Robin Foubister, Peter Lennard, Chris Rickards, Bruce Brookes.
Front Row: Warren Johns, Derek Cordes, Keith Mandeno, Noel Wilson, Dave Greenwood, Marcello Manzoni.

The idea of a reunion of the 1968-69 NZARP team had occasionally been floated by members at periodic Antarctic functions, the most recent being the 50th Anniversary of Scott Base held in Christchurch in November 2007 when some of us had afternoon tea with Sir Edmund Hillary at his last significant public appearance. Nothing more came of the idea for our own reunion until Graeme suggested it in an e-mail to Robin in late 2007. His suggestion was simply to capitalize on the round number of 40 and stage a reunion in October 2008 to symbolize the month when most of us journeyed south. This was the trigger, and after a flurry of messages to elicit expressions of interest and support from a few members of the old team, the 'SB40' (Scott Base 40th) reunion was all on.

Robin cobbled together a committee in early 2008 to organize the event. Christchurch was the obvious venue as it was the jumping off point for our polar adventure. As I was the only one known to live locally, I was nominated to arrange a place and time to hold the event. Although keen to contribute and happy to take on the role of convener and treasurer I was reluctant to do it alone so co-opted Hugh to assist with the initial planning of event activities. Robin and Keith in Auckland took on the massive task of locating and contacting everybody after 40 years of dispersal around the globe. Graeme was nominated to moderate on decisions, especially those affecting overseas attendees, and to help locate expedition members living abroad.

Lessons learned from earlier functions helped shape the program. Among them were the decisions to limit attendance to bona-fide expedition members or visitors to Scott Base in 1968-69 and to encourage wives to attend. We estimated we'd need to attract at least 60 guests to self-fund the reunion. Endless e-mails went around the world to track down lost expedition members. In spite of organisational ups and downs, our big reward came in being able to locate all the surviving members of our year at Scott Base and Vanda and having all but one attend. Wives and partners were especially encouraged to come. The 57 attendees included expedition members living in Canada, Cayman Islands, United States, and Australia as well as from all over New Zealand. The four man Italian Alpine Club Expedition we hosted in 1968-69 was represented by geologist Marcello Manzoni who shared photos and souvenirs of his team's Antarctic adventures that none of us had seen before.

Building on the theme of 'shared memories', I raided the Antarctic archives of the New Zealand government and local museums to supplement the collection of images being sent in by expedition members. I selected, printed and captioned 550 of the most representative photos of the 1968-69 expedition year and mounted them on 24 large philatelic display frames which were erected in one of the conference rooms at the reunion venue. Bruce Brookes helped to digitize many of the images from slides, and printed some of the photos. The conference room (appropriately called 'A' Hut after the original lounge and mess hut at Scott Base) was also used for members to display their personal Antarctic memorabilia. Extra display tables had to be brought in to accommodate all the photo albums, models of sledges, dogs and tractors, indoor game trophies, maps, flags and souvenir clothing. It was just a mass of wonderful memories generating endless chatter and laughter. For most of us, it was the first time we had been able to share such treasures with our old comrades.

To remind our visitors (and other guests at the hotel) that they were in Antarctic territory, a large "Welcome to the Gateway of the Antarctic" display sign was borrowed from the Christchurch International Airport and erected in the hotel foyer. It featured a panoramic view of life-sized penguins. The Antarctic theme dominated the hotel decor throughout the weekend.

Formal activities began on Friday evening (17 October 2008) with a cocktail evening in 'The Sledge Room' (our conference dining room named after the unofficial venue for "after work drinkies" at Scott Base). The room was appropriately decked out with a genuine manhauling sledge and flags of the three nations officially represented by the expedition in 1968-69. It was the time for everybody to 'meet and greet' and for wives to get to know everybody.

The following day we travelled by bus to the "Antarctic Attractions" at the Christchurch International Antarctic Centre. The visit included a pre-tour talk and viewing the live penguins at feeding time before visiting the main display galleries. For many, the tour highlight was the recently produced wide screen film featuring the Dry Valleys. The bus had everybody back in the city for lunch and a prearranged rendezvous at the Canterbury Museum. Group photos were arranged in the Museum Antarctic Hall, then a back of house tour took everybody downstairs to a room specially set up for our visit with specimens collected by the 1968-69 expedition. I gave a short talk about my 1995-96 restoration of Sir Vivian Fuchs' TransAntarctic Expedition Tucker Sno-cat Able that held special memories for most of the team. It is displayed alongside one of Ed Hillary's South Pole trip Ferguson tractors and together they form a prominent feature in the Antarctic display hall.

The keynote function of the reunion was the anniversary dinner on Saturday evening. Our two special guests, widows of our only deceased expedition members Nigel Millar and Doug Spence, honored us by cutting the anniversary cake modeled on a Scott Base hut. A small display with portrait photos had been set up in memory of the two men.

Sunday was free of organized activities. Some visitors attended nearby church services in the morning, and others took short tours of the city and surroundings, but most elected to gather in "A" Hut, study the items on display, and to talk. Bob Hancock, our expedition radio operator and assistant postmaster of 40 years ago looked after our own reunion post shop, selling genuine Ross Dependency stamps provided by New Zealand Post for the occasion, and souvenir postcards featuring expedition photos. A few people left to travel home on Sunday evening, and those who remained enjoyed a wonderful impromptu celebration dinner at which the remainder of the reunion wine was used up amid much laughter and singing, especially from Marcello with a memorable rendition of Volare and other Italian love songs.

It was a fitting end to the reunion that had turned out to be a highly charged and emotional affair for many who had lost contact but not forgotten each other. A few tears were shed on the final parting with the realization for some that this "SB40" reunion might be the last meeting with some friends of the old team. Proof of the success of the gathering has been evidenced by the renewed communication between Old Antarctic Explorers of 1968-69 who now regularly keep in touch.

Allan Guard of Fairlie
Base Engineer,
NZARP 1968-69
Christchurch, NZ,

GLOSSARY

ADELIE. The classic penguin in a tux. Found all round the Antarctic including large rookeries on Ross Island.

BIVOUAC. Temporary outdoor shelter

CONTRAILS. Clouds of condensed water vapour made by the exhaust of aircraft engines.

CREVASSE. A deep crack in a glacier

DRY VALLEYS. A row of snow-free valleys located in Victoria Land along the western coast of McMurdo Sound. The area is described as extreme desert.

CATERPILLAR D4. A small diesel bulldozer.

DSIR. New Zealand's Department of Scientific and Industrial Research.

FAST ICE. Sea ice that has frozen along the coast.

FERGUSON. A Massey Ferguson farm tractor fitted with tracks.

GLACIER. A river of ice often flowing down from mountain areas towards the sea.

GLACIER TONGUE. The extension of a glacier into the sea and usually afloat.

HERCULES. A workhorse aircraft known as the C-130. Its ability to land on flat areas almost anywhere was a boon to field parties.

IGY. International Geophysical Year 1956-58.

JARE. Japanese Antarctic Research Expedition.

KATABATIC WIND. The downhill wind that blows off the polar plateau of the Antarctic. It helped shape the Dry Valleys.

LEAD. A break or fracture in the sea (pack) ice. These vary in width and length and occur when the drifting ice breaks apart.

MORAINE. Rock debris deposited at the edges of a glacier.

MORSE CODE. A signaling system using dots, dashes and space. The common use was sending electric pulses along a wire. It was widely used up until a few years ago in aviation, at sea and in remote place. It is now mostly used by amateur radio operators.

MUKLUKS. A nice warm polar boot. Ours were bright yellow, nice and warm, with a felt liner. Soft sided, they looked like moon boots with the soles a good inch or two thick.

NEVE. The upper part of a glacier formed by granulated snow.

NZARP. New Zealand Antarctic Research Programme

NZBC. New Zealand Broadcasting Corporation

NZPA. New Zealand Press Association

PACK ICE. Floating sea ice packed together.

PIEDMONT. A low-lying glacier formed by glaciers flowing from mountains.

PRESSURE ICE. Ice squeezed and forced upwards by wind, tide and current. Good display of this at the tide crack near Scott Base.

QUONSET HUT. The name given to a semi-circular building.

SASTRUGI. Parallel wave-like ridges caused by wind on hard level snow.

SIKORSKY. US Navy Air Development Squadron (VX-6) helicopter used to transport personnel and for reconnaissance, construction support and light cargo hauling around McMurdo Sound and Ross Island. Based at McMurdo, the helicopter had a range of about 200 miles.

SKED. A slang term for radio schedule.

SKUA. Probably the southernmost of birds, this fellow is a scavenger, found all around the Antarctic continent and quite prolific in McMurdo Sound.

SNOW TRAC. A rubber tracked snow vehicle powered by an air-cooled Volkswagen engine.

SNO-CAT. Our Tucker Sno-Cat Able now resides in the Canterbury Museum Antarctic wing. Able was first used by Sir Vivian Fuchs for the TransAntarctic Expedition.

SUPER CONSTELLATION. A piston engine, propeller-driven military and civilian airliner that traces its heritage to World War 11 and Howard Hughes.

TAE. The 1955-58 Commonwealth TransAntarctic Expedition marking the first overland crossing of Antarctica via the South Pole.

TIDE CRACK. The tide rises and falls under the sea ice resulting in cracks, often many miles long and parallel to the shore.

VUWAE. Victoria University of Wellington Antarctic Expedition.

WANNIGAN. A small hut mounted on runners for shelter or storage and towed behind a tractor train.

WEDDELL SEAL. A mottled silvery brown seal that lives all round the Antarctic continent. They can grow to about 10 feet long and weigh 650-1000 pounds.

WILLIAMS FIELD. The United States Antarctic Program airfield at McMurdo that serves both the US and New Zealand efforts. It is approx. eight miles from McMurdo Station and is formed on almost 300 feet of ice floating over water.

WILLY WILLY. An Australian term for short-lived, intense whirlwinds or dust devils.

INDEX

MY ANTARCTIC LIBRARY

Antarctic Adventure, Sir Vivian Fuchs, Cassell and Company 1959
Antarctic Photographs, Ponting and Hurley, Jennie Boddington, McMillan London Ltd, 1979
Antarctic Psychology, Tony Taylor, Science Information Publishing Centre, DSIR, 1987
Antarctica, A.S. Helm and J.H.Miller, NZ Government Printer 1964
Antarctica, Kim Stanley Robinson, Voyager, 1997
Antarctica, Readers Digest, 1985
Antarctica, The Global Warning, Sebastian Copeland, The Five Tree Press.
Antartide, (Italian) Carlo Mauri, Zanichelli, 1968
Birds of the Antarctic, Edward Wilson, Blandford Press 1967
Call of the Ice, David Harrowfield, David Bateman, 2007
Captain Scott, Peter Brent, Saturday Review Press, 1974
Cherry, Sara Wheeler, Vintage 2002
Deepfreeze: On The Ice, Peter Clarke, Warren Krupshaw, Burdett and Company, 1966
Down To The Ice, Les B. Quartermain, NZ Govt. Printer, 1966
Hell With A Capital H, Katherine Lambert, Pimlico, 2002.
Improbable Eden, Bill Green, Craig Potton, Craig Potton Publishing 2003
Innocents in the Dry Valleys, Colin Bull, Alaska University Press, 2009
In a Crystal Land, Dean Beeby, University of Toronto Press.
Mawson's Will, Lennard Bickel, Stein and Day, 1977
Monday at McMurdo, David Burke, A.H. and A.W. Reed 1967 (novel
New Zealand and the Antarctic, Les B. Quartermain, NZ Govt. Printer 1971
Nothing Venture, Nothing Win, Sir Edmund Hillary, Hodder and Stoughton, 1975
Race to The End, Ross D.E.MacPhee, American Museum of Natral History.
Return to Antarctica, Adrian Raeside, John Wiley and Sons Canada Ltd, 2009
Scott of the Antarctic, Elspeth Huxley, Atheneum New York, 1977
Shackleton and the Antarctic Explorers, Gavin Mortimer, Carlton Books, 1999
Shackleton, Christopher Ralling, BBC Television Series, 1983
Shackleton's Boat Journey, Frank Worsley, Pimlico, 1999
Silas, Edited by Colin Bull and Pat F Wright, Ohio State University Press.
South From New Zealand, An Introduction, Les B. Quartermain, NZ Govt. Printer, 1964
South To The Pole, Les B. Quartermain, Oxford University Press 1967
South, Graham Billing, Guy Mannering, Hodder and Stoughton 1964, revised A.H. and A.W. Reed 1969
Sounds of Antarctica, Hank Curth, A.H. & A.W. Reed.
Terra Antarctica, William L Fox, Shoemaker and Hoard, 2005
Terra Incognita, Sara Wheeler, Random House, 1996
That First Antarctic Winter, Janet Crawford, South Lattitude Research Ltd, 1998
The Antarctic, H.G.R. King, Blandford Press, 1969
The Coldest Place on Earth, Robert Thomson, A.H. and A.W. Reed, 1969

The Endurance, Caroline Alexander, Alfred A Knopf, 2006
The Heart of the Great Alone, David Hempleman-Adams, Bloomsbury, 2009
The Worst Journey in the World, Apsley Cherry-Garrard, Carroll and Graf Edition, 1989
The Year of the Quiet Sun, Adrian Hayter, Hodder and Stoughton, 1968
Tom Crean, Michael Smith, The Mountaineers Books, 2000.

DVD
Ice, Marcus Lush, TVNZ.
Scott of the Antarctic, Synergy Archive Series.
Ernest Shackleton, To The End Of The Earth, Kultur International Films./
The Last Place On Earth, Masterpiece Theatre Presentation, Central Independent Television 1994.

NZARP

Scott Base 1968-69

AUTHOR BIO

Graeme Connell was born and raised in New Zealand and has spent most of his life writing, first as a newspaper journalist, editor and publisher in New Zealand, Fiji Islands, and Canada and later as a public affairs communicator with Mobil in New Zealand, Canada and the United States.

For the past decade he has owned and operated a commercial print business. His other adventures include cross country ski touring in the wilderness, cycle touring in New Zealand, the Canadian Rockies and Nova Scotia, He lives with his wife Lois in Alberta, Canada, and together they pursue their passion to photograph wildflowers in the short northern spring and summer.

A commemorative plaque in the pavement at the Captain Robert Falcon Scott statue in downtown Christchurch, New Zealand.
It reads: In recognition of the men and women who have assisted New Zealand's Antarctic Programme during the 50 years 1957-2007."